Ideological, Cultural, and Linguistic Roots of Educational Reforms to Address the Ecological Crisis

"The term 'specialist' is inappropriate to describe a writer whose work traverses so many intellectual boundaries. This collection represents some of the finest interdisciplinary scholarship available anywhere, and its importance is difficult to overstate."

—*David J. Flinders, Indiana University, Bloomington*

In this volume, C. A. (Chet) Bowers, whose pioneering work on education and environmental and sustainability issues is widely recognized and respected around the world, brings together a carefully curated selection of his seminal work on the ideological, cultural, and linguistic roots of the ecological crisis; misconceptions underlying modern consciousness; the cultural commons; a critique of technology; and educational reforms to address these pressing concerns.

In the World Library of Educationalists series, international scholars themselves compile career-long collections of what they judge to be their finest pieces—extracts from books, key articles, salient research findings, major theoretical and/practical contributions—so the world can read them in a single manageable volume. Readers will be able to follow the themes and strands of their work and see their contribution to the development of a field, as well as the development of the field itself. Contributors to the series include: Michael Apple, James A. Banks, Joel Spring, William F. Pinar, Stephen J. Ball, Elliot Eisner, Howard Gardner, John Gilbert, Ivor F. Goodson, and Peter Jarvis.

C. A. (Chet) Bowers taught at the University of Oregon and Portland State University, and was granted emeritus status in 1998.

World Library of Educationalists series

For more titles in this series visit www.routledge.com/World–Library–of–Educationalists/book-series/WORLDLIBEDU

From Practice to Praxis: A Reflexive Turn
The Selected Works of Susan Groundwater-Smith
Susan Groundwater-Smith

Learning, Development and Education: From Learning Theory to Education and Practice
The Selected Works of Knud Illeris
Knud Illeris

(Post)Critical Methodologies: The Science Possible After the Critiques
The Selected Works of Patti Lather
Patti Lather

Education, Ethnicity, Society and Global Change in Asia
The Selected Works of Gerard A. Postiglione
Gerard A. Postiglione

Leading Learning/Learning Leading: A Retrospective on a Life's Work
The Selected Works of Robert J. Starratt
Robert J. Starratt

Communicative Competence, Classroom Interaction, and Educational Equity
The Selected Works of Courtney B. Cazden
Courtney B. Cazden

Ideological, Cultural, and Linguistic Roots of Educational Reforms to Address the Ecological Crisis
The Selected Works of C. A. (Chet) Bowers
C.A. Bowers

Ideological, Cultural, and Linguistic Roots of Educational Reforms to Address the Ecological Crisis

The Selected Works of
C. A. (Chet) Bowers

C. A. Bowers

Routledge
Taylor & Francis Group

NEW YORK AND LONDON

First published 2018
by Routledge
605 Third Avenue, New York, NY 10017
4 Park Square, Milton Park, Abingdon, Oxon OX14 4RN

First issued in paperback 2023

Routledge is an imprint of the Taylor & Francis Group, an informa business

© 2018 Taylor & Francis

The right of Chet Bowers to be identified as author of this work has
been asserted by him in accordance with sections 77 and 78 of the
Copyright, Designs and Patents Act 1988.

All rights reserved. No part of this book may be reprinted or
reproduced or utilised in any form or by any electronic, mechanical,
or other means, now known or hereafter invented, including
photocopying and recording, or in any information storage or retrieval
system, without permission in writing from the publishers.

Trademark notice: Product or corporate names may be trademarks
or registered trademarks, and are used only for identification and
explanation without intent to infringe.

Library of Congress Cataloging-in-Publication Data
A catalog record for this book has been requested

ISBN: 978-1-03-257005-1 (pbk)
ISBN: 978-1-138-72262-0 (hbk)
ISBN: 978-1-315-19340-3 (ebk)

DOI: 10.4324/9781315193403

Typeset in Bembo
by Apex CoVantage, LLC

Publisher's Note
The publisher has gone to great lengths to ensure the quality of this
reprint but points out that some imperfections in the original copies
may be apparent.

Contents

Foreword

C. A. (Chet) Bowers (1935–2017) was an American environmentalist and educational scholar. Born in Portland, Oregon, Bowers received his Ph.D. from the University of California in 1962. His first academic position was at the University of Saskatchewan, where he taught for five years. In 1967, Bowers joined the faculty at the University of Oregon, where he wrote and taught for the next 26 years. In 1993, Bowers moved to Portland State University, retiring from that position in 1998. He continued to teach environmental studies in Oregon until 2012. Before his passing in 2017, Bowers maintained his prolific writing and speaking at universities in the United States, England, Canada, South Africa, China, South America, Australia, South Korea, Taiwan, and Japan. Bowers wrote 27 books and nearly 200 journal articles and book chapters on the cultural roots of today's ecological crisis. His writing has been translated into Spanish, Chinese, and Japanese.

The term "specialist" is inappropriate to describe a writer whose work traverses so many intellectual boundaries. What can be said is that Bowers has made major contributions in at least two fields: the foundations of education and environmental studies. Contributions in the foundations area can be traced back to Bowers' graduate work in social thought and Western intellectual history. Building on philosophical traditions, Bowers has developed a sophisticated critique of the industrial and today's digital ideologies of Western modernity and mass consumption. He has argued that modernity, rooted in the Cartesian thought of the Enlightenment, continues to provide the conceptual templates and habituated cultural practices that now overshoot the sustainability of natural systems.

In the area of environmental studies, Bowers' contributions have been twofold. First, he has challenged environmental scholars to think more deeply about the assumptions on which the field is based. In particular, Bowers has argued that common approaches to environmental education draw on the same root metaphors (change as progress, the individual as autonomous, and nature as a machine) that have marginalized indigenous cultures and fueled the globalization of Western influence and cultural colonization. To put this argument in broader perspective, Bowers has challenged the ideal of the autonomous individual as the nucleus of rational thought. It is this conception of rationality that has been attacked more generally by contemporary feminist scholars as an illusion that only glorifies the privilege of men, by behavioral economists as a poor predictor of decision-making, and by cognitive and evolutionary scientists as neglecting the reliance of individuals on patterns of thought and affect that are inherited from the past.

Bowers' second contribution to environmental studies includes the development of an ecological perspective that returns to the original meanings of the term "ecology" from the Greek *oikos*, a word initially used to describe home and household practices, their cultural ordering, and the interdependent relationships among family and community members. Ecologies, from this perspective, include not only the interactions among living creatures and their environment, but also the cultural patterns, cooperative agreements, conceptual feedback systems, and all of the shared interests that sustain community life. By the late 1980s, Bowers had come to refer to these ecologies collectively as "the cultural commons." Broadly viewed, the commons movement is often seen today as a countervailing force against the excesses of free-market capitalism. The source domain for the commons is historically associated with places (parks, libraries, airways, and public lands) that belong to an entire community. Bowers has extended this metaphor to encompass the shared knowledge found in a wide range of creative works, including the arts, folkways, agriculture, scientific research, and public education.

Bowers has developed these contributions through a number of lines of inquiry. Much of his early work drew on the sociology of knowledge. This tradition, revived in the 1960s by writers such as Peter L. Berger and Thomas Luckman, is centrally concerned with the social construction of reality and the influences of prevailing ideas on social institutions. Bowers would later include the work of Michel Foucault in this tradition as well. Seeking to understand thought within its social context, sociologists of knowledge draw on the twin concepts of socialization and culture. The latter, culture, may be the most fundamental to Bowers' scholarship.

Bowers' conception of culture has been largely consistent across his career. Drawing on both sociology and sociolinguistics, he has emphasized two primary functions of culture. First, culture provides the architecture for social and psychological processes, doing so by way of shared knowledge, skills, values, beliefs, understandings of temporal and spatial relations, attitudes, traditions, power relations, and material objects. As the totality of learned behavior, culture constitutes the historically transmitted patterns of understanding that allow us to deal with experiences not as unique but as integrated with symbolic and interpretive systems. In this sense, culture is the medium through which ideas and beliefs develop, analogous to the growth of bacteria or cells in a nutrient-rich substance. For Bowers, the development of thought through culture is also closely linked to communication and language. Like many sociolinguists, he sees culture as communication and vice versa.

The second aspect of culture prominent in Bowers' work is that it functions largely at an assumed or taken-for-granted level of understanding.

Culture may seem like it is simply absorbed because its implicit, implied, and connotative meanings do most of the work. Everyday experiences depend on their assumed meanings. A person reading an English newspaper, for example, recognizes that they are reading English rather than, say French or Japanese. Yet, this "recognition" is not typically made at any conscious level of awareness. Watching a film in a cinema, to take another example, requires the suspension of disbelief. We simply take this as one of the conventions of drama, without needing to willfully remind ourselves that "this is a fiction" or "this is a film."

In working through the implicit nature of culture, Bowers has drawn significantly on the social sciences, including Erving Goffman's work on frame analysis. For Goffman (1974), frames provide the templates for the social organization of experience. Frames answer the question, "What is going on here?" Returning to an earlier example, going to the cinema can be framed as entertainment, a date, a family night out, and so forth. All of this framing is done with little reflective thought or rational scrutiny as long as others share the same frame. Because such meanings are assumed, anthropologists sometimes speak of culture as invisible or hidden. But what "hides" culture is its pervasiveness and our own tacit familiarity with cultural norms. As William Wordsworth put it, "the world is too much with us."

Bowers has used frame analysis and a range of other theories to deepen our understandings of how culture shapes worldviews, social traditions, and the meanings of lived experience. Sociological inquiry, however, is not the touchstone of Bowers' work. Instead, social inquiry serves the normative and ethical purpose of promoting the types of cultural change that increase ecological sustainability. Bowers has developed this ecological perspective by building on the work of deep ecologists (e.g., Wendell Berry), eco-justice activists (e.g., Vandana Shiva), and ecologically minded anthropologists. Among the latter group, Bowers has given particularly close attention to Gregory Bateson's (1972) work on "the ecology of mind." Freudian psychology has sought to understand the mind by delving within its subconscious systems and by analyzing components of the system (such as the id, ego, and super-ego). Bateson, a semiotic anthropologist, extended the mind outward as fully integrated with the physical, cultural, and social environment of which the mind is simply a part.

Bateson's ecology of mind is a cultural ecology of ideas, norms, traditions, and shared patterns of thought. Much like natural ecologies, cultural ecologies are looped systems of information exchange and transformation. In Bowers' 2011 book on Bateson's ideas, he further developed this perspective in order to critique the Western notions of the autonomous individual that underlie supposedly progressive educational reforms. Bowers has argued that what we typically think of as the individual or "self" is defined by its membership in multiple self-correcting systems of exchange and interdependence. As such, the self is not bounded by our skin-encapsulated bodies. As cognitive scientists Steven Slowman and Philip Fernbach (2017, pp. 14–15) put it,

"Our skulls may delimit the frontier of our brains, but they do not delimit the frontier of our knowledge."

A closely related ecological principle, also fundamental to Bowers' work, is that no part of an ecology exercises unilateral control over the other parts. Understanding ecologies as interdependent asks that we think of causality as circular rather than linear. To take an everyday example, we might describe a wall-mounted thermostat as a device that controls the temperature of a room. Yet from an ecological point of view of information exchanges, the thermostat is part of a looped system in which the temperature in the room "controls" the thermostat, which controls the furnace (or AC), which controls the temperature. The components in the system are not controlling, but responsive. We can extend this circuit by adding other systems such as the self or the power plant (coal-burning where I live). Bateson (1972) points to the significance of this principle in stating:

> When you narrow down your epistemology and act on the premise "What interests me is me, or my organization, or my species," you chop off consideration of other loops of the loop structure. You forget that the eco-mental system called Lake Erie is part of *your* wider eco-mental system—and that if Lake Erie is driven insane, its insanity is incorporated in the larger system of *your* thought and experience.
>
> (p. 484, emphasis in original)

Bowers' cogent development of ecological principles, together with the centrality of culture and the questioning of our intellectual history, are reflected throughout this collection. Bowers selected writings for this volume that represent his work in four areas: language, the cultural commons, ecological intelligence, and critiques of technology. In Part I, Bowers reminds us that words have a history and that the historical meanings of words reproduce the cultural root metaphors that continue to shape contemporary patterns of thought that undermine the sustainability of natural systems. The chapters in this part examine language in three critical contexts, including metaphor, environmental education, and how print promotes abstract thinking. Part II examines the cultural commons, a form of sustainable wealth that includes knowledge and skills that are shared through intergenerational mentoring, observations, the arts, cultural narratives, and community traditions. Part III focuses on ecological intelligence, how such forms of intelligence enable groups to live within the limits of their bioregion, and why ecological intelligence requires a radical shift in thinking that goes beyond individual self-interest and political power. In Part IV, Bowers argues that the digital revolution and the globalization of computer-mediated learning are now having a profound impact on world cultures by reinforcing the basic ontological misconceptions that we live in a static, objective world while also marginalizing forms of knowledge based on emergent relationships and interdependence.

Overall, this collection represents some of the finest interdisciplinary scholarship available anywhere, and its importance is difficult to overstate. While Bowers offers an intellectually incisive and sometimes scathing critique of seemingly progressive educational philosophies, digital technologies, Western ideologies, and rampant consumerism, his work is not without its optimistic side. Throughout Bowers' career, he remained devoted to education as a process of cultural change. The forms of cultural change that he advocates require a deep understanding of our cultural interdependence as well as why and how information, including data, is neither culturally nor environmentally neutral. Moving toward sustainability requires understanding that information is learned not from objects or abstract ideas, but from emergent and dynamic relationships. Where relationships strain or weaken a community's sustainability, Bowers has urged us to seek out their cultural underpinnings and find correctives to the cultural narratives that give rise to these unsustainable relationships to begin with. Through education it is possible to mediate or recover the sustainable relationships necessary to stem the ongoing ecological destruction that we are witnessing on a global scale.

David J. Flinders
Indiana University, Bloomington

References

Bateson, G. 1972. *Steps to an ecology of mind.* New York: Ballantine.
Bowers, C. A. 2011. *Perspectives on the ideas of Gregory Bateson, ecological intelligence, and educational reforms.* Eugene, OR: Eco-Justice Press.
Goffman, E. 1974. *Frame analysis.* New York: Harper Colophon.
Slowman, S. and Fernbach, P. 2017. *The knowledge illusion.* New York: Riverhead Books.

About the Author

C. A. (Chet) Bowers taught at the University of Oregon and Portland State University, and was granted emeritus status in 1998. In retirement he was still active professionally, writing and giving invited talks at international conferences that focused on educational reforms promoting a more sustainable future. He was invited to speak at 42 universities in the United States, including Harvard and Stanford Universities. Invitations to speak at 41 universities in other parts of the world included the University of Trondheim, University of Zagreb, University of Queensland, University of Cape Town, Rhodes University, York University, University of Toronto, the Chinese University of Hong Kong, Trinity College (Dublin), Universidad Catolica Boliviana San Pablo, Swiss Foundation for Environmental Education, Seoul National University, and National Chung Hising University (Taiwan). Notably, he was asked by Vice-President Al Gore to be the featured speaker at a dinner/seminar (held at the Gore residence) on the influence of metaphorical thinking on environmental and technology policies. He was one of six Western scholars, along with the former Chinese Minister of Culture, invited to speak at the International Congress on Culture and Humanity in the New Millennium, sponsored by the government of Hong Kong and the Chinese University of Hong Kong. Bowers gave the John Dewey Memorial Lecture in 1982, and in 1991 was voted the most outstanding environmental faculty at the University of Oregon. In June 2016 he gave an invited talk at a major international conference on educational reforms that address the ecological crisis.

Credits

The following chapters have been previously published and their publication details are given below.

Chapter 2: The Cultural Aspects of the Ecological Crisis

C. A. Bowers, *Education, Cultural Myths, and the Ecological Crisis: Toward Deep Changes*, New York: State University of New York Press, 1992, pp. 9–33.

Chapter 4: Language Issues that Should be the Central Focus in Teacher Education and Curriculum Studies

Adapted from C. A. Bowers, *In the Grip of the Past: Educational Reforms that Address What Should be Changed and What Should be Conserved*, Oregon: Eco-Justice Press, 2013, pp. 41–62.

Chapter 5: Why the George Lakoff and Mark Johnson Theory of Metaphorical Thinking Fails to Address the Linguistic Issues Related to the Ecological Crisis

C. A. Bowers, *University Reform in an Era of Global Warming*, Oregon: Eco-Justice Press, 2011, pp. 141–155.

Chapter 7: The Political Economy of the Cultural Commons and the Nature of Sustainable Wealth

C. A. Bowers, *The Way Forward: Educational Reforms that Focus on the Cultural Commons and the Linguistic Roots of the Ecological/Cultural Crises*, Oregon: Eco-Justice Press, 2012, pp. 27–45.

Chapter 8: The Cultural Mediating Role of the Professor—Across the Disciplines

C. A. Bowers, *University Reform in an Era of Global Warming*, Oregon: Eco-Justice Press, 2011, pp. 93–112.

Chapter 10: The Challenge Facing Educational Reformers: Making the Transition from Individual to Ecological Intelligence

C. A. Bowers, *The Way Forward: Educational Reforms that Focus on the Cultural Commons and the Linguistic Roots of the Ecological/Cultural Crises*, Oregon: Eco-Justice Press, 2012, pp. 15–26.

Chapter 11: Gregory Bateson's Contribution to Understanding Ecological Intelligence

C. A. Bowers, *Perspectives on the Ideas of Gregory Bateson, Ecological Intelligence, and Educational Reforms*, Oregon: Eco-Justice Press, 2011, pp. 13–40.

Chapter 12: Rethinking Social Justice Issues Within an Eco-Justice Conceptual and Moral Framework

C. A. Bowers, *In the Grip of the Past: Educational Reforms that Address What Should be Changed and What Should be Conserved*, Oregon: Eco-Justice Press, 2013, pp. 131–152.

Chapter 17: Is the Digital Revolution Sowing the Seeds of a Techno-Fascist Future?

C. A. Bowers, *Reforming Higher Education in an Era of Ecological Crisis and Growing Digital Insecurity*, Minnesota: Process Century Press, 2016, pp. 149–163.

1 Where to Start in Addressing the Cultural/Linguistic Roots of the Ecological Crisis

Challenging the Misconceptions
Underlying Modern Consciousness

The problem we all face is how language continually misrepresents the world in which we live. This is not new in human history, but today if we do not wake up to the misconceptions encoded in the languaging processes we take for granted, the consequences will gradually evolve from being inconvenient, to being disruptive of our taken-for-granted worlds, and then to social unrest that will end in violence as each person and social group seeks to survive the catastrophic changes taking place in the Earth's natural systems. There is a growing awareness of how the biases and misconceptions extending back hundreds (even thousands) of years encoded in the meaning of words have influenced the patterns of discrimination in social relationships and determined who has been able to exercise power over others. But how the vocabularies inherited from the past also misrepresent key characteristics of the world in which we live, that is, the ontology that lies beyond the different cultural explanations of what is "real," needs to be recognized if there is any hope of educational reforms addressing the deep cultural roots of the ecological crisis.

So what are the basic changes that need to be made in how we understand reality? The following represent a partial list of misconceptions widely shared within the dominant Anglo-American culture:

1   As Thich Nhat Hanh put it, "Nothing can exist by itself alone" (2002, 47). That is, everything from rocks, organisms, humans, ideas, values, and cultural artifacts is part of an emergent, relational, and co-dependent world. The impermanence that characterizes all aspects of the world, as well as the information exchanges (the varied semiotic systems) within the ecologies of relationships, means that nothing can be fully and finally explained.

2   The beliefs about universal ideas and values, as well as the idea that individuals are autonomous thinkers, are long-standing misconceptions in the West. For a universal to exist it cannot be expressed in a culturally influenced language—as the language (including mathematics, which is always interpreted in terms of social meanings) cannot help but reproduce the culture's past misconceptions and silences. In short,

cultural languages can only provide interpretations that are too often misunderstood as objective accounts.

3   The basic units of organization are the cultural and natural ecological systems, and not the supposed self-forming and self-governing individual—or other autonomous entities such as plants and animals. The ecological systems can be divided into natural ecologies and cultural ecologies as long as the latter are understood as dependent upon the self-renewing capacity of the natural ecologies. All ecological systems have a history that influences their current and future prospects, with some being destructive of life-forming and sustaining processes.

4   There can be ecologies of identity, language, economic systems, cultural commons, and so forth. There are also ecologies of exploitive practices, narratives based on misconceptions, and willful ignorance. Using ecology as a conceptual framework brings into focus the co-dependent, semiotic, mutual support, as well as the historical roots of what otherwise might be understood as an autonomous entity, event, idea, plant, animal, individual, conflict, achievement, and so forth.

5   Most of a person's cultural knowledge is acquired and experienced as part of her/his taken-for-granted world. It is largely shared with others at this same taken-for-granted level of awareness through the ongoing cultural ecologies of communication. Naming what is otherwise taken for granted often leads to making it explicit, thus enabling the person to reflect on how it is reinforced by other taken-for-granted cultural patterns and even to introducing changes that may affect the deepest cultural assumptions.

6   The idea that change, especially scientific and technologically based changes, leads to progress is a myth that too often results from the failure to consider the importance of the traditions that are being lost—such as the loss of privacy, the loss of work that enhances the person's skill and the need to engage in mutually supportive activities, and intergenerational knowledge and skills that reduce dependence on ecologically destructive consumerism.

7   The historical conceptual/moral influences encoded in the vocabularies whose meanings are largely taken for granted, which are the basis of thought and communication, may be made explicit in ways that lead to changes. That we are dependent upon languaging processes, including the ability to adopt new analogs that reframe the meaning of words, means that we live in an interpreted world—which in turn means that objective knowledge is a widely held misconception that sets in motion different power relationships. Even the facts, such as the date when Lincoln was assassinated and the start of World War I, are interpreted in ways influenced by the interpreter's taken-for-granted assumptions. The writer, reader, speaker, and listener who might claim the objective status of knowledge and data all rely upon an inherited language that

frames the process of interpretation in ways that too often reproduce cultural assumptions taken for granted in the past.

8   Ecologies are characterized by the patterns that connect, by interdependencies, and information (semiotic) networks. In effect, everything communicates and exercises the intelligence unique to its species and culture. For humans, the exercise of ecological intelligence requires being aware of the ways in which language—spoken, written, and encoded in the built environment—carries forward the analogs settled upon in the past. Revising the meaning of words involves identifying new analogs that reflect the current awareness of what is being communicated through both the natural and cultural ecologies. It also involves mindful awareness of the differences that make a difference in the behaviors within both cultural and natural ecologies—and that lead to different responses being communicated through the networks of communication.

The educational reform implications of these ontological realities of how we exist in the world means that students need to learn the many ways in which their culture reinforces the old mythologies about objective knowledge, abstract thinking, and technological innovations as inherently progressive forces, and that traditions inhibit progress and individual self-expression. They also need to recognize the cultural ways of thinking and practices that overturn traditions, including community sustaining traditions misrepresented as sources of backwardness, supposedly universal ideas such as free markets, the need to colonize other cultures, and so forth.

The educational process should thus become more of a dialogue guided by questions that lead students to consider the cultural patterns and interdependencies that are a taken-for-granted part of their daily experiences. This, in turn, allows students to reflect on and discuss what needs to be intergenerationally renewed and what needs to be changed. In effect, the curriculum should shift more toward making explicit the cultural patterns and interdependencies of the students' culture that would otherwise be taken for granted. The moral guidelines should take account of what strengthens communities of mutual support and that reduce the adverse impact on natural systems.

Textbooks and computer programs may be a source of useful information and even challenging as problem-solving games. However, they cannot reproduce the information networks and the taken-for-granted cultural patterns that not only constitute the students' life world, but also limits their future prospects. The reason for this limitation is that which is taken for granted cannot be digitized—except as yet another taken-for-granted pattern. And, secondly, the education of the people who write the learning programs seldom understand their own culture as well as other cultures as ecologies nested within natural ecologies. The result is that what students encounter in a computer-mediated curriculum is the reduction of the lived culture to what appears in print or visual reproductions of historical events that are often lacking in an in-depth understanding of local contexts.

The book I wrote in the early 1970s (Bowers, 1974) focused on the need to make explicit the taken-for-granted cultural pattern that was based on the misconceptions of earlier generations who were unaware of environmental limits. It also stressed that science and technology would be unable to shield humans from the consequences of changing the chemistry and self-renewal capacity of natural systems. Everything I have written since then has made addressing how to reduce the human impact on the Earth's natural systems my primary concern.

As humans are cultural beings shaped in unconscious ways by the language systems that pass forward the deep and generally taken-for-granted assumptions of the culture, I have focused on how so much of the process of primary socialization, where students are learning something for the first time, involves passing forward the misconceptions, silences, and prejudices (and occasional wisdom) of previous generations. This phase of socialization occurs when the student is most vulnerable to adopting the taken-for-granted understandings of her/his significant others. Thus, the early and most vulnerable phase of joining the conceptual/linguistic community is the most critical as many of the culture's core assumptions are intergenerationally passed forward—with many of the assumptions, such as thinking of the world as constituted by things, plants, animals, other autonomous individuals, progress, and so forth, left unquestioned in later stages of formal education.

The following essays highlight how the changes in the self-renewing capacities of natural systems, such as extreme changes due to global warming, expanding populations that exceed local natural resources, the rise in social tensions, and the combination of spreading poverty and police-state levels of surveillance, make it imperative that classroom teachers and university professors become aware of the cultural myths that now underlie the modern, consumer-dependent culture that is accelerating the rate of ecological collapse.

This process of primary socialization, depending upon the level of awareness of the significant others, has the potential of leading to an empowered form of primary socialization where students are encouraged to make explicit the discrepancies between the meaning of words that have been framed by the analogs settled upon by earlier generations, such as how the potential of women was understood, the idea of objective knowledge promoted within the scientific community, and the complexity of the everyday cultural patterns that are too often ignored by students (and others) as they align their daily lives with the abstract and stereotypical ideas they acquired from others. Both my critiques of how education at all levels continues to pass forward the misconceptions and silences of earlier eras, as well as how to enable students to reflect on whether the explanations they acquire from others enables them to recognize the differences between ecologically sustainable and unsustainable patterns, has made becoming aware of one's own taken-for-granted cultural patterns the primary focus. This of course also requires becoming aware of what others, particularly those in positions of power, take for granted.

All of the following essays focus on the linguistic/ideologically driven processes that reinforce the taken-for-granted world of previous generations, as well as how to juxtapose older ways of thinking with the current evidence of environmental destruction and educational reforms that strengthen the self-sufficiency of communities that enable people to live less consumer-dependent lifestyles. All of the essays start with a background summary of the degradation of nature systems, followed by a critical analysis of cultural patterns that have generally been ignored by educational reformers who address various social justice issues but ignore the deepening ecological crises that change everything, such as how an over-reliance upon print-based cultural storage and thinking undermines the exercise of ecological intelligence and how digital technologies are driven by a liberation ideology whose roots can be traced back to such abstract theorists as John Locke and Adam Smith. Most of the essays also discuss how ecologically sustainable patterns of thinking and values can be introduced in the classrooms.

Scientific studies have shown changes in different ecosystems, such as the melting of the ice on Greenland that has the potential to raise the level of the world's oceans by 20 feet, and the changes in the pH levels of the world's oceans that are limiting the ability of calcifying organisms to reproduce themselves. Moreover, the digital revolution is leading to more algorithms replacing humans in work settings and has made privacy a thing of the past. In this context, the more thoughtful students who are beginning to consider the life they want to lead as adults are likely to transition between rage and depression as they consider the environmental and cultural legacy left them by their parents and by earlier generations who put self-interest and the myth of progress above the well-being of future generations.

The myth of progress, in effect, enabled people to ignore the evidence of how rapidly the Earth's natural systems were being degraded—and to assume that progress has an ontological standing, like the myth of heaven and the equally mythic thinking of scientists who gave us the industrial and now digital revolution, along with the ability to genetically engineer the future direction of evolution itself. Unfortunately, the life-threatening changes in natural systems, as well as the digital technologies that are putting us on the road to being subjects in a society organized by corporate and police-state algorithms, will be the realities for which today's youth will hold their parents' generation accountable.

Many of the essays are focused on helping students to recognize the cultural traditions and practices that are sources of hope—even as the collapse of ecosystems increases the level of social stress and chaos, which we are now witnessing with millions of refugees moving across national borders. The source of hope, that is, the local cultural commons that need to be further organized in ways that perform the role of the guild systems of the Middle Ages that were centers of mutual support, are also models of resistance to the consumer-dependent market system now being globalized in the false belief that science, technology, and capitalism can ensure progress even as ecological systems collapse.

References

Bowers, C. A., 1974. *Cultural Literacy for Freedom*. Eugene, OR: Elan Publishers.
Thich Nhat Hanh 2002. *No Death, No Fear*. New York: Riverhead Books.

2 The Cultural Aspects of the Ecological Crisis

A cultural theme most middle class Americans have grown up with is that change is not only a normal aspect of existence, but that it is also progressive. The steady stream of technological innovations, from computers that exploit dimensions of time that exceed what most people can comprehend to advances in biomedicine, suggest that this aspect of our belief system is still intact. Other aspects of society appear to be breaking down—urban violence, teen pregnancies, the spread of drugs, increasing numbers of homeless and impoverished, skyrocketing national debt. Yet science continues to promise technological and thus human advancement. Change continues, but a growing body of evidence strongly suggests that our sense of living in one of the most progressive times in human history may be an illusion. The computer, for example, may be increasing our efficiency in finding solutions to certain technical problems, including how to store and manipulate the mountains of information we have come to believe necessary for effective decision making. But this form of change pales in significance when compared to changes taking place in the life sustaining capacity of our habitat.

The increasing number of reports on the changing characteristics of the planet's ecosystems indicate that over the long term they are in decline. They also suggest that in the immediate decades ahead we face the growing prospect of having to change the most fundamental aspects of our belief system and patterns of social life. The challenge will be to see through the illusions of a consumer-oriented, technologically based existence, to alter the premises upon which the belief system of the dominant culture is based, and to retain those aspects of our past cultural achievements that are compatible with a culture in equilibrium with the carrying capacity of the natural systems that make up the biosphere.

There are many dangers. One scenario of future possibilities is a continuation of the state of self-absorption that characterizes the addictive personality (and culture) until the consequences of environmental disruption lead to economic dislocation and a significant loss of human life. There are other possible scenarios, one extreme being the emergence of a fascist form of government that attempts to regulate human life according to master plans drawn up by technocratically oriented bureaucrats. The other extreme would

be a widely held sense of despair and futility about the prospects of changing cultural beliefs and practices in time to avert overshooting the carrying
capacity of the natural systems. With the long-term prospects appearing so
bleak, people may turn to the short-term pleasures and unconcerned attitudes
associated with consumerism.

Although the rate of change in the ecosystems has been accelerating
in the last hundred years, the real acceleration occurred in the post-World-
War-II era. Awareness within the dominant culture of human interdependence with natural systems is a relatively recent phenomena. Aldo Leopold's
gentle and poetic *A Sand County Almanac* provided an analogue for living
in a less exploitive relationship with the other forms of life that make up
the environment. His book, calling for the development of an ethic toward
the land that would supplant the tradition of viewing the environment
primarily in economic and political terms, was published in 1946. Rachel
Carson's *Silent Spring* (1962) provided in a way that could be understood by
the general public the scientific evidence that our technological approach
to the environment threatened the basis of all forms of life—including our
own. By the time of Paul Ehrlich's *The Population Bomb* (1968) and Barry
Commoner's *The Closing Circle* (1971), the environmental movement was
gaining a wider following within certain sections of society. The point being
made here is that the recognition that cultural practices cannot evolve inde-
pendently of a concern for the well-being of the habitat goes back a mere
forty years; and it was an awareness unevenly understood and appreciated
within mainstream society. Native American cultures, of course, had evolved
in ecologically responsive ways; but what could have been learned from their
thousands of years of experience in adapting to the unique characteristics of
their habitat was ignored because they were perceived as unenlightened and
pre-modern.

The discrepancy between the view of cultural change held by the middle
class and the nature of environmental change can be seen in the findings of
scientists who are studying the interactive environmental systems that make
life possible as we know it. The schema of understanding based on Western
culture represents change as a progressive expansion of human possibilities:
personal freedom and individual advancement, control over life-threatening
situations, power to solve problems and direct the course of future events,
and of course an expansion of possibilities for consumption. The expansion
of human population (from 1.6 billion in 1900 to 5 billion in 1986) and the
corresponding increase in world economic activity (from 0.6 trillion dollars in 1900 to 13.1 trillion in 1986) are indeed impressive figures. But this
growth in the number of people and the scale of economic activity has also
increased the disruptive impact of humans on the habitat. Here the trend line
indicates a decline in the viability of natural systems.

The cultural image of progress is, in part, based on scientific-technological
developments, such as the synthesizing and introduction into the environment of approximately seventy thousand different kinds of chemicals, as

well as the widespread use of technologies for transportation that burn fossil fuel—to cite just two examples. The scale of human impact on environmental systems, which scientists view as accelerating over the last three hundred years—with a particularly large jump in the rate of change occurring during the last 50 years—is indeed daunting. According to the lead article in a *Scientific American* special issue (September 1989), "Managing Planet Earth," the planet since the beginning of the eighteenth century has lost forest cover equal in area to the size of Europe. And the rate of deforestation has now increased to between 70,000 and 110,000 square kilometers of land a year, which is equivalent to the combined land mass of the Netherlands and Switzerland. To put it another way, an additional one percent of the total forest cover is being lost each year, and the rate is accelerating. The concentrations in the atmosphere of methane and carbon dioxide—two chemicals viewed as contributing to the green-house effect—are estimated to have increased 25 percent over the last 300 years. Taking into account changes in current levels of population and economic activity, it is projected that the concentration of carbon dioxide in the atmosphere will have doubled sometime after the year 2030—which will be well within the lifetime of the children of the readers of this book. Although there is not agreement among scientists about the factors that contribute to the greenhouse effect—the influence of cloud cover, the exact inventory of the earth's carbon dioxide absorbing biomass, and so forth—there is agreement that the earth's atmosphere is warming at an accelerating rate, and that this trend will have unknown consequences on precipitation patterns, agricultural production, forest zones, and—given even just the expected rise in sea level—on human settlement.

Changes in other natural systems also show a clear downward spiral in life-sustaining capabilities. Essential sources of water, such as aquifers, are being rapidly deleted in certain parts of the world, and major rivers such as the Nile, Ganges, Mississippi, and Colorado are running at reduced flow rates as ever more demands are placed on them. The levels of acidification of lakes and soil are increasing dramatically in the Northeastern part of the United States and Canada, as well as in Europe, China, and parts of India. With more land being brought under irrigation, salinization, estimated at twenty percent of irrigated land in the United States, is becoming a problem. Other changes in the life sustaining capacities of the environment include the impact of human waste and chemicals on marine ecosystems, the accelerating of species extinction as deforestation and other land use practices reduce the amount of natural habitat, and the dumping on the environment of toxic waste (over twenty billion pounds in the United States in a single year) as well as other human garbage.

Although scientists and others studying the changes in the earth's ecosystems may disagree on the figures used to understand the interactive patterns that sustain the biosphere as a living system, and the passage of time will make obsolete much of the data currently used to understand the scope of the crisis, the direction of environmental change is unmistakable. With the possibility of the earth's population doubling within the next thirty years from 5 to 10 billion people, most of the increase will occur in Third World countries that are trying to increase their levels of economic activity in order to accommodate the increase in population and to raise living standards, the rate of environmental degradation will likely increase. The 1987 report of the World Commission on Environment and Development, *Our Common Future,* estimates that the anticipated increases in population may result in a five- to ten-fold increase in what is now a 13-trillion-dollar world economy. This may be seen as continued evidence of human progress when viewed through our cultural framework but in terms of further impact on already stressed ecosystems it has catastrophic implications.

This brief overview is not meant to be alarmist; it simply summarizes trends that are documented on a daily basis in newspapers with headlines that announce environmental problems around the world, from "Italy Copes with Summer Slime" (125 miles of coastline strangled in a mass of algae that stretches 20 miles out to sea) and "East Germany's Ghost Towns" (towns that are disappearing from the map because of practices surrounding the strip mining of low-grade, high-sulfur brown coal) to "LA Making Last-Gasp Effort to Clear the Air." Television coverage of environmental "disasters" and "catastrophes" are becoming so frequent that these words are losing their power to hold the public's attention; ozone holes, oil spills (even on the scale of the Exxon Valdez), burning and cutting the rain forest, and nuclear contamination seem like an unending succession of media events that for a brief period occupy the public's attention before being displaced by new announcements and revelations. This overview thus is related to what the scientific reports and media coverage are not dealing with—namely, the cultural and, by extension, educational aspects of the problem.

The special reports on the environment, which range from the Worldwatch Institute's yearly publication, *State of the World,* to such journals as *The Economist* and *Scientific American,* frame the problem in terms of a rationalistic approach to problem solving. This is not surprising, because the people who either carry out the studies of environmental change or summarize the data for a nonspecialist audience have largely been educated at universities in an ideology that holds human action to be based on a rational process. According to this view, information is the basis of rational thought. Thus, the reports are framed in a way that will provide rational people with the data necessary for understanding the nature of the problem, and for acting, primarily through enactment of new environmental legislation, in a more environmentally responsible manner.

The report of The World Commission on Environment and Development, *Our Common Future,* and the Worldwatch Institute's *State of the World* are typical of the way in which the environmental crisis is being presented. Both present scientific data and suggest critical areas in which new policies and legislation—both on the national and international level—are needed to guide human behavior toward a way that is more environmentally sustainable. Both reports refer to humans in terms of people, society, and governments; but references to culture (indeed, cultures) are totally lacking. The special issue of *Scientific American* on "Managing Planet Earth" contains an article "The Changing Climate," where this rationalist view of human behavior is clearly represented. Reflecting on the social significance of recent studies of climate change, Stephen H. Schneider, a leading scientist in the area of climatology, commented that,

> I am often asked whether I am pessimistic because it will be impossible to avert some global change. At this stage, it appears, no plausible policies are likely to prevent the world from warming a degree or two. Actually I see a positive aspect: the possibility that a slight but manifest global warming, coupled with the larger threat forecast in computer models, may catalyze international cooperation to achieve environmentally sustainable development, marked by stabilized population and the proliferation of energy-efficient and environmentally safe technologies.[1]

William D. Ruckelshaus, writing in the same issue of *Scientific American,* urges that "a clear set of values consistent with the consciousness of sustainability" be articulated by national leaders.[2] Provide the data, state the issues clearly, utilize government incentives, and people will have a rational basis for changing their life styles. Again, we find the assumption that humans are rational beings when they have the right data, but no acknowledgement that people are essentially cultural beings, that the world is made of multiple cultures, and that culture makes the outcome of the political process far more problematic than is recognized by people who hold a rationalistic point of view.

Since current approaches to framing the ecological crisis are conditioning us to accept the rationalist approach to problem solving, they help to insure that the human dimensions of the crisis are never really understood at the deepest levels. The argument here is not against being rational; rather the main issue is an overly narrow view of the well-spring of human thought and behavior. The other problem with the rationalist approach is that it ignores how different cultural groups organize their way of understanding on fundamentally different assumptions and root metaphors; thus the human aspect of the ecological crisis is not simply a matter of people, societies, and nation states—those misleading metaphors of Western colonialism—but of cultural differences that cannot be easily reconciled or changed by using a political process based on the Western forms of rationalism so evident in environmental reports. The cross-cultural dimensions of the ecological crisis, while exceedingly important and complicated, are not however the main focus of this work.

Our main concern here is with the middle class culture which exerts such a dominant influence in American society, and with how the belief system of this group, which underlies so many environmentally disruptive practices, is perpetuated in the public schools and universities. Pronouncements on the necessity of other cultural groups' changing their environmentally destructive practices may help inflate our moral sense of superiority and bolster our self-image as ecologically responsible citizens. But as largely ritual behavior this diverts attention from the part of the problem that we can actually do something about. Directing our energies to bringing our own society's dominant culture into closer balance with the long-term sustaining capacities of our own environment is justified because we, along with the other Western industrialized, consumer-oriented societies, are major contributors to the problem. The technologies that support our life styles deplete nonrenewable resources and contribute to multiple forms of pollution on a scale vastly disproportionate to our percentage of the world population.

Focusing on our own situation should not, however, be taken to mean that what other cultures do about their relationship is no concern of ours. If we have learned anything in recent years it is that the ecosystems of the planet are interactive with each other. Thus, destroying the forests in one part of the world ultimately will have an influence on the precipitation patterns and changes in soil conditions in other parts of the world. Overproduction of food grains in our country, which involve the use of techniques damaging to our water supplies and soil, alter the economics in other countries with more marginal soils—forcing them to adopt even more disruptive practices. Other examples of the interactive nature of human practices and ecosystems could easily be cited. But the most immediate challenge is to address at the level of formal education our own disruptive cultural patterns—while keeping an eye on the influence that our practices and policies have on the manner in which other cultural groups use the world's commons of soil, oceans, forests, and atmosphere.

In recent years a wide variety of groups have taken on the mission of changing attitudes and social practices relating to the environment. They range from environmentalists who argue for a more responsible form of stewardship of the environment to advocates of "deep ecology." The latter is more an umbrella term covering a wide array of groups who view the ecological crisis as raising fundamental questions about our belief and value systems. The more serious "deep ecology" thinkers, such as Arne Naess, Warwick Fox, and Alan Drengson, have attempted to articulate the philosophical basis for an ecologically grounded view of self. But the American tendency toward syncretism is fully evident in the attempts of others to find the path to an ecologically enlightened form of existence in such diverse traditions as Taoism, Buddhism, libertarianism, feminist spirituality, communitarianism, Greek mythology, Christianity, Shamanistic cultures, and so forth. Often the impression is left that the pathway to ecological balance is in embracing as many of these traditions as possible.

Though somewhat less ebullient about the transformation of consciousness, the Greens are perhaps the most important social and political movement to have emerged in response to the growing deterioration of our habitat. The Greens in North America share the more comprehensive political agenda that has made them so highly visible in Western Europe and Australia. As the following statement makes plain, they see the need for a complete restructuring of society to restore the system to a smaller scale where the skills, interests, and responsibility of the individual matters again. To quote from the position paper of the Green Party of the Federal Republic of Germany:

> A complete restructuring of our current near-sighted planning is necessary. We consider it mistaken to believe that our present spendthrift economy can still promote human happiness and the fulfillment of life goals. Just the opposite occurs. People have become more harried and less free. Only to the extent that we free ourselves from an overdependency on a materialistic standard of life, and make individual self-realization possible again, and recognize the limits of our own inner nature, will our creative powers be able to free themselves to form life anew on an ecological basis.[3]

Their concern with participative involvement, nonhierarchical social structures, and an ecologically and socially oriented economy, to cite just a few of their proposals for changing the basic direction of Western modernization, also involves specific proposals for changes in education. The organization of schools and a curriculum that fosters class divisions and stratified bodies of knowledge essential to furthering a technocratic society are specifically rejected.

They propose that the old competitively based model of education be replaced by an approach to education that emphasizes the following content:

- practical training for both teachers and students in handicrafts, industry, and agriculture, in order to reduce the gap between education and the working world.
- school children able to learn outside of the school, and real life situations brought into the schools. The separation between school time and leisure time must be eliminated. Music, theater, painting, work and play, all must find their place in the school.
- the encouragement of thinking in terms of interrelated systems as the ongoing goal of teaching, in order to encourage a better understanding of social interrelations, ecological cycles, and prevailing contradictions. The school should be able to train the student to recognize the interests which lie behind social and personal conflicts. They should be taught to solve conflicts peacefully between people, to formulate their own concerns, and, together with others, to find effective ways of expressing these interests.[4]

Beyond the environmentalists' recommendations on the importance of teaching recycling in schools and avoiding hazardous waste, as well as other common-sense practices that accompany the annual "Earth Day" observances, the Greens are one of the few groups to address the problem of educational reform from an ecological perspective. With the exception of their recommendation that students be taught to think in terms of interrelated systems, their proposals, however, appear to be little more than the restatement of much of the educational thinking of the 1960s, with its emphasis on individual self-realization in an open, nonauthoritarian classroom. The Greens' more coherent vision of how different aspects of society—work and technology, energy, agriculture, zoning and community development, education, and so forth—must be reorganized in order to decentralize decision making, and to replace the traditional anthropocentrism with biocentrism, reframes the goals to be served by these educational reforms.

But the orientation of the Greens, like the ideological traditions on which they draw so heavily, reflects the contradictions and tensions of these earlier ways of thinking—as well as their silences. The Greens' ideal of personal autonomy and their view of the nature of freedom raise questions about how these traditional political ideals are to be reconciled with the Greens' nonanthropocentric view of the human relationship to the natural world. Whether the other aspects of their political agenda for transforming a technocratically oriented society into cooperatively based, smallscale communities can be attained through the democratic process is also problematic. But the aspect of their thinking, as well as that of the environmentalists and most deep ecologists, that is of most concern here is the failure to consider the influence culture has on human thought and behavior.

Technology, profits, competition, multinational corporations, Taoism, and feminist spirituality are all aspects of culture. In fact, anything that humans do, including those activities and artifacts that survive over time, can be understood as representing the influence of culture. Because it is so encompassing, varied in its expression, and formative in its influence, culture is less amenable to the political process than the Greens and other environmentally conscious groups recognize. This does not mean that culture and the political process can ever be domains of human activity distinct from each other: the forms of politics are also grounded in cultural patterns and codes. And the political process—if it leads to change—involves changes within other dimensions of the culture. Recent political change brought about by the feminist movement demonstrates this point. But the lack of specific focus on the cultural aspects of the ecological crisis seems to limit remedial action to those political arenas where disagreements rooted in economic interests and ideological differences have made change exceedingly difficult and slow. Witness the time it takes to enact environmental legislation, and the compromises that must be made.

Attempts to maintain a sustainable habitat through the political process—whether in the form of demonstrations, working in the hallways and committee rooms of the legislature, or spiking trees—effectively preclude utilizing the full potential of the classroom to help ameliorate the crisis. There is no single cause for any aspect of the ecological crisis, but there are complex and interconnected cultural patterns, beliefs, and values that collectively help to introduce perturbations into ecosystems, causing them to go into decline. To put this another way, there is no single cultural cause for cutting old growth forests in the Northwest or for engineering automobiles as status symbols rather than for fuel efficiency. Practices and beliefs far from the scene of the ecological crime—so to speak—are contributing through often invisible vectors; for example, our attitudes toward packaging and reading newspapers have something to do with cutting old growth forests. It is not simply a matter of capitalism, as some extremists might argue.

The changes in some practices—like dumping toxic wastes, deforestation, depleting fisheries, and losing topsoil and ground water—require immediate attention, and this will mean utilizing the political process to enact legislation. Attempting to effect changes in these areas through the long-term process of educationally guided cultural changes would be too slow a process. But the long-term aspects of learning to live in ecological balance also require giving attention to those aspects of culture that will have an influence on the taken-for-granted beliefs, values, and social practices that people will hold in the future. To use the ongoing feminist movement as an example, it is possible to see how direct political action is being translated into legislation, and how the more long-term cultural adjustments are being worked out in the context of the classrooms as students learn to think about work in less gender-biased language and engage in other behavioral patterns that treat people on a more equitable basis. The two processes—political and educational—are going on simultaneously; while they complement each other, the processes have distinct characteristics that must be taken into account.

To consider the broader and longer-term process of bringing mainstream culture into a sustainable balance with the habitat, it is necessary that the complexity of culture must be recognized as well as why it is so difficult to be aware of this complexity. Rather than fit the discussion of how educators should respond to the ecological crisis into the controversial frameworks of the deep ecologists, Greens, and environmentalists—whose disagreements can become excuses for others' indifference—I want to propose a different approach. The focus on culture, which is the medium of the classroom, allows for more educators to become involved, and at different levels. It also addresses the symbolic dimensions of a long-term solution—which has to do with the cultural patterns, beliefs, and values that will be part of the taken-for-granted attitudes of future generations. Before we take up the more direct educational implications of this approach, it is necessary to focus more directly on the nature of culture itself and on why certain aspects of the dominant Western culture have made the formative influence of culture even more difficult to recognize. Although public school and university education, in both their curricular content and their patterns of teaching, are cultural processes, we have not really understood the special educational issues raised by the culture-language-thought connection. One of the reasons for this is connected with the specific cultural pattern of thinking now being brought into question by the ecological crisis. Thus, a deeper understanding of culture, as well as the specific cultural patterns now being recognized as problematic, may also help guide us toward a more a ecologically responsive approach to public school and university education.

Clifford Geertz's definition of culture provides a good starting point for illuminating one of the reasons culture is so difficult to recognize. The other reason, which has to do with the specific set of beliefs and assumptions that has had a privileged position in Western thought for nearly 400 years, will be taken up later. For Geertz, culture can be understood as the shared patterns that set the "tone, character, and quality" of people's lives—ranging from what is viewed as food, how it is prepared and eaten, to the categories used to understand the world and the human beings placed in it. The following explanation by Geertz points to the pervasive nature of culture in human experience: "Culture patterns—religious, philosophical, aesthetic, scientific, ideological—are 'programs'; they provide a template or blueprint for the organization of social and psychological processes, much as genetic systems provide such a template for the organization of organic processes."[5] Of course, we must recognize here that Geertz is representing culture in a highly metaphorical manner; unlike the chemical code of a gene that regulates the form that life will take, the culture that provides the patterns guiding human experience can be partly understood at the explicit level of awareness and changed as a result of thought—which may even include changes based on misconceptions. But the key point in Geertz's explanation is that the patterns that make up a culture are the largely invisible yet always present sources of authority in people's lives. Even the power of rational thought and creativity are expressed in terms of culturally shared patterns.

This brings us to a second critically important feature of culture, namely, that the patterns used by people as the basis for their experience—which range from the body's message system for communicating interpersonal relationships to the categories used to organize thought—are experienced as part of the person's natural attitudes. A way of understanding a person's attitudes is to view them as what the person takes for granted in experience. As we are here identifying an aspect of culturally based experience of which by definition, we are not aware, it may be useful to cite examples that we can recognize as the taken-for-granted foundations of our own experience. The practice of associating "up" with good and "down" with bad, adjusting inter-personal spatial distance according to social context and status relationships, designing a house, and deciding who is included in our family and commu-nity are just a few examples from the everyday world. Patterns of belief and of social interaction that are felt as immutable may be problematic in fact, as when cultural patterns of the early European explorers led them to view the New World as an economic resource to be staked out for exploitation. More recently, we have witnessed the damaging effects of technological innova-tions that were not anticipated because of fundamental attitudes about the progressive and ameliorative nature of new technologies. Beliefs and social practices may take a positive form, like the patterns we follow in successfully interacting with others, driving a car, and hiking on a mountain trail. It is not the unexamined nature of cultural patterns that is the problem, but the patterns themselves. The invisible nature of cultural patterns becomes a problem only when they are the source of undesirable consequences, like the cultural practice of designing automobiles to require greater use of air conditioning that results in the release of ozone-damaging chemicals into the atmosphere.

There is another dimension of given patterns that needs to be recognized. The patterns we unconsciously re-enact can be understood as lived traditions. That is, the cultural patterns—ranging from what is viewed as constituting wealth, how we organize and use space, to our sense of time, to identify just a few examples—represent past forms of understanding that have been encoded in the patterns that underlie current experience. As other members of our cultural group share these patterns of encoded knowledge, meaningful communication and social interaction is possible. If all these patterns were made explicit, judged, and improvised on a purely individualistic basis, communication would become impossible. In effect, making explicit all patterns, if it were possible, would contribute to a condition of nihilism where the authority of traditional patterns would be relativized, and values and commitment would vary in accordance with personal opinion. The critical issue here, related to Geertz's view of cultural patterns, is that even the person who is questioning all forms of authority in people's lives is being unconsciously influenced (even driven?) by cultural patterns. But this does not mean that all cultural patterns evolve in a way whereby only existentially meaningful and socially useful traditions are preserved and the outmoded ones are made explicit and so reconstituted in a way that in turn becomes the basis of experience for the next generation. There are also cultural patterns that continue to have authority in our lives even though they threaten our existence over the long-term.

For example, the distinctive set of beliefs that many people in the dominant culture associate with progress and modernity fosters mental habits particularly unsuited for recognizing either implicitly assumed beliefs or evidence that challenges the idea that change is inherently progressive. These beliefs, born in a period of self-conscious emancipation from what was viewed as the "Dark Ages," were seen as the basis of human empowerment: in overcoming illness and death, the drudgery of work, the suppression of human freedom, and the barriers to material success. The partial fulfillment of the promise has not been entirely illusory, but the essential core of these beliefs evolved in a way that lost touch with the realities of an environment that has natural limits. To make this point in a different way, the hubris of this mind set assumed that environmental limits could be transcended through the resourcefulness of human rationality.

Key aspects of this belief system still provide a basis for understanding the implications of the changes taking place in the environment. A culturally specific view of individualism, of the rational, and of the nature of language can be traced directly to the seventeenth- and eighteenth-century thinkers who laid the conceptual foundations for the evolution of modern Western consciousness. While often disagreeing on important issues relating to the nature and source of ideas, the founding fathers of modern consciousness left a distinct and still obvious legacy. John Locke (1632–1704) helped to establish the primacy of the individual by arguing that the individual is to be understood as existing prior to society: that is, as a biological entity living in a "state of nature. Membership in society came about as a result of the realization that the self-interest of the individual could not be pursued on a more rational (predictable) basis in the state of nature where there was no third party who could adjudicate disputes.

René Descartes (1596–1650) also started with the individual who, through a process of radical doubt, supposedly is then able to exercise a form of rationality free of the influence of both tradition and culture. Descartes' formulation of thought as an inner mental process helped establish today's dualisms of mind and body and of thought and nature. The contemporary understanding of objective knowledge, procedural thinking, and the reduction of a thing or process to its smallest components also have their roots in Descartes' epistemology.

Ernst Cassirer, in summarizing the influence of Descartes on subsequent Enlightenment thinkers, observed that "reason is now looked upon rather as an acquisition than a heritage. It is not the treasury of the mind in which truth like a minted coin lies stored; it is rather the original intellectual force which guides the discovery and determination of truth."[6] Furthermore, as the rational process was considered to be everywhere the same, involving a particular relationship between reflection and sense data, there was no need to account for cultural differences. The idea that language might encode the thought patterns of a cultural group, thus influencing the "rational process," was likewise unimaginable to these early founders of modern consciousness.

In the course of our discussion, the terms "Cartesianism" and the "Enlightenment" will be used as a means of identifying the historical and cultural origins of ideas and values that now tend to be associated with a universal form of modernism. "Cartesianism" will be used when referring to those aspects of modern consciousness that can most directly be traced back to Descartes' mode of thinking, with its dualisms and linear procedures of thinking. Although Descartes' legacy has undergone important modifications, we are still, at the deepest level of our thought process, Cartesian thinkers. This can most easly be seen in the curriculum and the teaching styles that characterize the educational process from the early grades through graduate school.

The term "Enlightenment" will be used to designate the somewhat later emphasis given to the authority of reason in guiding people's lives, the belief in the inevitability of progress and, as Alexander Pope so succinctly put it, the belief that "the proper study of mankind is man." The two traditions represent a distinct cultural pathway that now seems to be increasingly problematic. Our task will be to illuminate how current interpretations of key ideas, assumptions, and values associated with modern consciousness are putting us in a double-bind, particularly in our response to the ecological crisis.

Currently, the way individualism is understood within the dominant culture varies in terms of the past socialization of different subgroups. Academics and technical experts who write on environmental issues would tend to emphasize a different, though not incompatible, set of attributes; for them, individualism would be more associated with rational self-determination than the expressive form of individualism to which many other groups subscribe. Keeping in mind that "individualism" is a metaphor that encodes different peoples' various associations, examples, and analogues of what it means to be an individual, it seems safe to say that individualism is generally associated with the idea of freedom. But "freedom" is also a metaphor that encodes different schemas of understanding, depending upon the historically formative analogues. But the most powerful analogue, which is that of the autonomous individual, suggests that freedom is a matter of choosing one's own values, one's self-identity, and future. Within the cultural mainstream, differences seem to arise more in terms of how to achieve the fullest expression of individual freedom, rather than over the deeper questions associated with freedom itself. Some argue that rationality is the basis of individual authority while others argue for a more emotive basis of individual authenticity.

In *Habits of the Heart,* Robert Bellah and his colleagues write eloquently about the consequences of making the individual self the basic social unit, and question whether this modern image, cut off from communal involvement and responsibility, can sustain either a meaningful public or private life.[7] But there are other consequences of this view of individualism that relate more to the foundations of the myth of the autonomous, self-directing individual. One consequence is that thinking of self as an autonomous individual hides the multiple dependencies upon patterns of thinking, use of technologies, and reenactment of social conventions that have been handed down from the past. To put this another way, the current image of individualism (which Edward Shils points out as being part of a Western tradition of thought) disconnects the "individual" from tradition at the level of self-understanding. But as the wheel of thought does not always have to impinge on the road of everyday reality, individuals (in being absolutely dependent upon tradition for coping with every aspect of daily life) are left in a schizophrenic condition where their view of freedom contradicts their reenactment of traditional patterns and practices. For our purposes, the important point here is that the current image of individualism does not recognize the complex nature of tradition and the authority that it has in people's lives. This is, as we shall later see, a critically important issue in any serious discussion of the characteristics of an ecologically sustainable culture.

Another consequence of associating individualism with freedom is that it prevents a deeper awareness of the dimensions of experience as influenced by culture. As discussed earlier, the way cultural patterns are taken for granted and thus not part of the person's self-awareness, helps maintain the myth of individual autonomy. One aspect of culture put out of focus by the Western emphasis on the self-directed individual is how language, with its roots deep in the past, influences thought and behavior. Language can be understood as encoding the thought processes (actually, the mental ecology) of earlier stages in our cultural history. For example, thinking of the heart as a pump encodes the earlier assumptions about the mechanistic nature of the universe; thinking of North America as "the New World" encodes the privileged European perspective (many native peoples referred to the continent as "Turtle Island"); and thinking of creativity as an original act of the individual encodes assumptions that evolved along with the Western view of modern art. What the current view of individualism obscures is that language provides the important schemata or conceptual frameworks that guide the thought process of the individual. In effect, the patterns of individual thought are culturally rooted; while this means that there is far less original thought than is now proclaimed, it does not mean that thought is entirely determined by the encoding characteristics of language. As demonstrated here, it is possible to make the underlying patterns explicit and to reconceptualize them. This may lead to minor changes in the collective conceptual mapping process. Or, as attempts to shift away from the Cartesian mind set suggest, the changes may be more profound and reach deeper into levels of cultural practice.

This brief discussion of the relationship between language and thought, where language is more deterministic in direct relation to the individual's taking its cultural formulations for granted, is related to how the myth of individual autonomy contributes to thinking about personal responsibility. The conceptual schema (what we can now recognize as part of a culturally and historically specific way of thinking) that leads to thinking of self at the same time as both self-directing and the center of an autonomous rational and moral authority, undermines the sense of being interdependent with the larger social and biotic community. Responsibility is thus viewed in terms of self-interest; and if there is any awareness of living in an interdependent world it is likely to be viewed as an unwelcomed constraint on individual freedom. The discussion of language, tradition, and beliefs (which are all different aspects of the cultural milieu that makes human life possible) points to a basic fact that is not recognized by the modern form of consciousness: namely, that, as Gary Snyder points out, life involves participation in information and food networks.

The German philosopher Martin Heidegger argues that the dualisms that reflect Descartes' influence on Western consciousness not only include separating mind from body and the mind from the external world, but also a particular way of thinking about the world. According to Heidegger, the fundamental change in our way of thinking introduced by Descartes is to conceive and grasp the world as a picture.[8] This sense of being an observer who can make separate and objective judgments strengthened other Western cultural assumptions that extend even further back in time. One of these is the anthropocentric view of the world; that is, the world is to be understood and valued only from the perspective of human needs, interests, and sense of rationality. This positioning of "man" at the apex or center of the world, depending upon which tradition of Western thought you follow, has had the effect of privileging humans as superior to other life forms by virtue of their distinctive capabilities as rational beings.

But the form of rationalism we have created over the last four hundred years of industrial development in the West has problematic characteristics other than those related just to the spectator/anthropocentric way of understanding the "external" world. When compared with the patterns of thinking among traditional peoples, which have been described as consensual in that the members reflect and act within a shared overarching conceptual framework that recognizes many traditional forms of authority, the modern rationalist thinker can be more easily recognized as operating according to a different set of norms.[9] Alvin Gouldner, the late political sociologist, has identified how these norms are based on a model that is individualistic and competitive. These norms or "rules of critical discourse" include: the justification of assertions, the use of evidence rather than the invoking of traditional forms of authority (those who represent the authority of tradition, sacred texts, communal memory, and so forth), the voluntary standing of listeners, and the competitive nature of the forum in which assertions are defended.[10] In more popular terms we have referred to these norms as providing for John Stuart Mill's open marketplace of ideas, and as a competitive arena where truth emerges because of the preponderance of evidence—only to be challenged by alternative interpretations or the emergence of new evidence.

This view of the rational process, in being based on a competitive model that locates the authority in the rational process of individuals, rejects tacit and more contextually grounded forms of knowledge. In turn, the outcome of this more context-free form of thinking is viewed as the basis for making judgments that apply universally; that is, the outcome of this rational process is regarded as valid regardless of cultural context or time frame—until overturned by a newer way of understanding.

There are several characteristics of this way of thinking that help insure the privileged standing that it now has in the Western educational process, including many parts of the world where modernization is being based on the Western model. The first has to do with the emphasis that this model of the rational processes places on presenting its claims in a supposedly open, competitive arena where argument and evidence are the basis for establishing truth claims. People educated to think theoretically (that is, to think abstractly) and who have the economic and political means to establish the evidence to support their theory are going to prevail over those people who do not possess the elaborated speech codes necessary for playing by the norms of the competitive model. Forms of authority that are grounded in experience or serve as keystones of their symbolic universe are not recognized as credible by those who uphold the norms that privilege their own patterns of rationality. A second characteristic has to do with the dynamism of this approach to rationalism. We have heard on countless occasions of the advantages for humankind from the "free marketplace of ideas"; indeed, many of the advances in technology, institutional safeguards of civil liberties, and social opportunities have their origins in the openness of the rational process.

But this dynamism has another dimension that relates directly to the environmental crisis we are addressing here, and it is this dimension that has largely escaped the attention of people who give speeches and write articles and books extolling the connection between the modern view of the rational process and human progress. The continual introduction of new knowledge, and the mechanical, social, and political technologies that quickly follow, involve a continual process of cultural experimentation. That is, the innovations alter the patterns and relationships that characterize the fabric of social life. But the mind set that is oriented toward the introduction of ever-newer thinking and technology does not give much attention to the social consequences of the previous cultural experiment—unless of course the consequences, such as breakdown of a nuclear reactor or deformities resulting from a particular drug, are especially dramatic and socially visible. This tendency toward cultural experimentation is such an important and complex issue that we shall return to it when we address the implications of the ecological crisis for such long-standing ideals as academic freedom.

Reading about why our own belief system is so difficult to recognize is not nearly as exciting as being told that the moral relativism spreading across America can be traced to the influence of a couple of German philosophers or that our educational test scores have dropped to the bottom of the list for Western industrialized societies. But the hidden nature of belief systems, our own as well as those of other cultures, becomes a matter of great importance when it comes to the political challenges we face in the immediate years ahead in bringing about the changes on an international level in governmental economic policies and individual life styles necessary for reversing the disruptive effect of humans on the environment. In the past, the use or threat of military force was a basic aspect of the political process. The use of rational persuasion was also an important aspect of international politics. With the growing realization of the catastrophic dangers (and economic costs) associated with achieving political ends through military action we seem now to be left with the rationalist approach that leads to seemingly endless negotiation in order to bring about, for all the effort, only incremental changes in governmental policy. Examples of the slowness of governments to act in the face of serious environmental problems easily come to mind: the failure to address the root causes of the acid rain problem that affects Northeast Canada and the United States, the inability to address the problem of drift-net fishing in the Pacific, to cite just two examples. The international political process, in both environmentally constructive and destructive ways, is also being influenced through an increasingly internationalized media. Television coverage seems to be awakening people in living rooms around the world to the decimation of dolphins, elephants, and rain forests, and fouling of shorelines. But this grassroots political process is counterbalanced, if not overwhelmed, by the power of the media to indoctrinate people with the message that to be modern means to adopt an increasingly materialistic life style.

Aside from economic issues, which are very real, one of the reasons the international political process does not work in a way that fits the Western rationalist's expectations is the culture factor. The hiddenness of culturally specific assumptions, which influence how the constructs of another cultural group about reality, continually gets in the way of the rational process that is often upheld as the universal standard by Westerners. Ironically, it turns out that one of the least exciting things to read and think about—the constituent elements of our own belief system and why these particular beliefs cause us to ignore the cultural dimensions of the ecological crisis—is perhaps one of the most important in addressing the crisis.

We would be falling into the rationalist trap if we accepted the idea that understanding cultural differences will finally enable us to settle differences on a rational basis. The foundations of cultural belief are not rationally formulated, and thus are too deeply rooted in the person's interconnected psychology, language, and map of the world. Although more limited than rationalists would like (as perhaps a defense against the fascist tendencies in the rationalist's position), increasing cross cultural understanding is essential to facilitating communication about how issues are understood, and thus why peoples of different cultures frame the issues in such diverse and often irreconcilable ways. The solution to this problem is admittedly long-term and, again, will be only partial.

But there is a dimension to this problem of understanding how culturally based ways of thinking and acting have an adverse effect on the environment that is closer to home. This dimension has to do with changing the foundations of our own belief system, and while it may seem in the eyes of some as more defensible than setting out to change the belief systems of other cultural groups, it is no less difficult. As the ongoing feminist movement has shown, as well as other recent social changes that could easily be cited, it is possible to affect changes in people's thought and behavioral patterns at the deepest levels when the changes become a new part of taken for granted culture. And like the feminist movement, part of whose energy was focused on the belief system taught in the schools, we will direct our attention to this critical area of cultural reproduction.

Suggesting that the public schools and universities should be one of the arenas for reconstituting the conceptual foundations of a culture that is eco- logically out of balance is likely to elicit the response from many educators that schools simply cannot take on any more social responsibilities. A response might be to sink further into a state of deep despair, as schools have become so politicized that they have achieved few of the social goals envisaged by social reformers. I myself vacillate between these two responses, but I cannot see that the public schools can be excluded from the crisis. The double bind of expanding cultural demands on a contracting resource base is not the sort of issue that will go away. Self-denying strategies, failure to exercise intelligence commensurate with the scope of the problem, and a diverting of energy to other political agendas (even such heavily contested ones as issues related to race, gender, and social class) will not displace the ecological crisis as the most pressing political and moral issue facing humankind.

The failure of current educational reformers, both conservative and technicist, will be taken up in the next chapter; but two of the more salient reasons for not excluding public education from the process of cultural renewal will be laid out here. Public schools and universities are only two of many institutions that pass on the culture, and they do not have the ability to influence thought and expectation to the same degree as the media. Nevertheless, they exert an important influence on what aspects of the culture are transmitted to the next generation and in determining which aspects will be understood at the explicit level of awareness and what will be part of the person's stock of unexamined "knowledge." A second reason our institutions of formal education have a special importance in the ecological crisis is that the socialization of students involves encountering in a more systematic way the language and conceptual frameworks that underpin the mainstream culture. If this conceptual framework, which also includes a sense of the moral order, is derived primarily from the period of history that produced the Industrial Revolution, we face the tragic prospect of the next generation being caught in a conceptual double bind where the ability to understand the problem will be dependent upon the same patterns of thought partly responsible for the scope of the crisis.

As the world's population begins to experience the environmental consequences of moving toward the 10 billion mark, along with the effects of economic activity associated with creating additional jobs and raising the standard of living (read: dumping more toxic waste, depleting and contaminating dwindling supplies of fresh water, and so forth), the ecological crisis will become a concern of television executives, theologians, governmental officials, and even the heads of multinational corporations. Each group, in terms of the aspects of the culture they most influence, will be faced with rethinking the most fundamental aspects of their taken-for-granted belief system. This also applies to educators—in both public schools and the universities.

Notes

1 Stephen H. Schneider. "The Changing Climate." *Scientific American,* vol. 261, no. 1 (September 1989), p. 78.
2 William D. Ruckelshaus, "Toward a Sustainable World." *Scientific American,* vol. 261, no. 1 (September 1989), p. 169.
3 *Die Grünen: The Program of the Green Party of the Federal Republic of Germany.* Bonn: Die Grünen, 1989. p. 4.
4 *Die Grünen,* p. 40.
5 Clifford Geertz. *The Interpretation of Cultures.* New York: Basic Books, 1972, p. 261.
6 Ernst Cassirer. *The Philosophy of the Enlightenment.* Boston: Beacon, 1951, p. 13.

7 Robert N. Bellah, Richard Madsen, William M. Sullivan, Ann Swidler, and Steven M. Tipton. *Habits of the Heart: Individualism and Community in American Life.* Berkeley and Los Angeles: University of California Press, 1985.

8 Martin Heidegger. *The Question Concerning Technology and Other Essays.* New York: Harper Colophon Books, 1977, p. 129.

9 Robin Horton. "Tradition and Modernity Revisited." in *Rationality and Relativism,* Martin Hollis and Steven Lukes, eds. Cambridge, MA: MIT Press, 1982, p. 39.

10 Alvin W. Gouldner. *The Future of Intellectuals and the Rise of the New Class.* New York: Seabury Press, 1979, pp. 28–29.

Part I

An Ecological Paradigm for Understanding Language Issues

3 The Relational and Emergent Nature of Cultural and Natural Ecologies

Charlene Spretnak (2011, p. 1) goes to the heart of the problem when she writes that "Our hypermodern societies currently possess only a kindergarten-level understanding of the deeply relational nature of reality." For all our technological and intellectual achievements, we have missed, as she puts it, "the way the world works" (p. 1). As our everyday lives are dependent upon awareness of what is being communicated through the relationships of which we are aware (the car speeding in the wrong lane; the non-verbal communication of the Other, use of a word that encodes a prejudice, and so forth), Spretnak's statement points to a complex cultural double bind: that is, how print and now what can be digitized lead to ways of thinking that misrepresent the emergent and co-dependent world we live in as fixed and made up of autonomous entities. The challenge here is to provide a conceptual framework for understanding how computer-mediated learning, which still relies on print as the principal means of communication, perpetuates students' conceptual misunderstandings and thus limits their awareness that all aspects of life are emergent, relational, and co-dependent. As will be explained more fully, there are no autonomous entities except in the world constituted by print, English nouns, and the misconceptions passed forward in public schools and universities. Unfortunately, this linguistically constructed world of reifications, such as free markets, freedom, data, rational thought, and so forth, has been imposed on the dynamic life-forming and sustaining processes.

How Print Misrepresents Life Processes

Print has had special standing since the invention of the printing press. Books, maps, treaties, and newspapers have been acclaimed as contributing to democracy and a literate public. But there is a downside to print that brings into question whether this old technology is, on balance, capable of representing the ecological challenges we face in the twenty-first century. The following summary of the characteristics of print needs to be considered, especially now that more of the students' learning is mediated by computers that rely upon the technology of print. The often-ignored characteristics of print include the following: 1) Print provides only a surface knowledge of an event, process, and context; 2) What is encoded in print quickly becomes dated and thus misrepresents the relational and emergent processes in the different cultural and environmental ecologies; 3) Print reinforces the misconception of providing an objective account; 4) Print lends itself to being reified and treated as having universal validity; 5) The impression of objectivity associated with printed accounts is further reinforced when the conduit (sender/receiver) view of communication is adopted; 6) Although print can be used to provide an historical account and even a description of contexts, too often print is used in ways that hide that words have a history; 7) The combination of print and the conduit (that is, the sender/receiver) view of language undermines awareness that most words are metaphors, and thus have a history. Print also privileges sight as the primary basis of knowing, while excluding reliance upon the other senses as sources of information about what is being communicated through the relational world we call ecologies.

In effect, the transition to computer-mediated learning, which allows for the use of other media, continues the dominant tradition in the West of marginalizing an awareness that there are no objects, ideas, facts, data, individuals, or events that have not been influenced by their relationships within larger and more complex ecologies that have a history, and that interact with other ecological systems—both natural; and cultural.

The Paradigm Shift that is Underway

In order to understand this criticism, it is first necessary to provide an overview of how the paradigm that emphasized a mechanistic view of organic processes, of individual autonomy in a human-centered world, and of science and technology leading to endless progress and material abundance, is now being challenged. The primary importance of these challenges, beyond providing a more accurate understanding of life-forming processes, is that it provides the conceptual framework necessary for addressing how to live more ecologically sustainable lives.

What does Spretnak mean by referring to the world as relational, and why do the print-based misconceptions become especially important as the world's population expands toward the nine billion mark, along with a consumer lifestyle that is further undermining the life-sustaining capacity of natural systems? The answer to both questions can be traced to a single word: Ecology. In the middle of the nineteenth century, this word represented what has become the modern translation of the early Greek word *oikos*, which supposedly referred to the management of the Greek household. I say "supposedly" as the translation by the German biologist, Ernst Haeckel (1834–1919), was accepted within the scientific community of that day as a fact. This example of metaphorical thinking, where the management of the environment was understood as like the management of the household, led to a very narrow understanding of ecology as the study of the behavior of natural systems. Lost in translation was what Haeckel, as an early proponent of Darwin's theory of evolution, was less able to understand. Namely, that for the early Greeks, oikos encompassed the norms governing a wide range of cultural practices.

This science-dominated understanding of ecology is now beginning to change. A small group of scientists is developing the new field of biosemiotics that expands understanding of how the word ecology moves us closer to understanding the emergent nature of life processes. There are now increasing references to the ecology of identity, the ecology of language, the ecology of bad ideas, the ecology of colonization, the ecology of marriage, and so forth. That the explanatory power of the word ecology can be applied to any aspect of the natural and cultural world, as well as to how they interact, is based on the recognition that ecology is another word for co-dependent relationships, and the multiple patterns of communication that are integral to all relationships.

This is where the thinking of Alfred North Whitehead, Gregory Bateson, the biosemiotic-oriented scientist, Charlene Spretnak, and other linguistic and anthropological thinkers such as Clifford Geertz, Walter Ong, and Richard E. Nisbett becomes helpful. Nisbett's *The Geography of Thought: How Asians and Westerners Think Differently . . . and Why* (2003) is especially useful as it clarifies how the languages in East Asia rooted in Confucianism, Taoism, and Buddhism focus awareness on the world of relationships, and the moral codes that should guide these relationships.

For example, the relational orientation of Confucianism can be seen in its fivefold guiding principles: *Jen* "involves simultaneously a feeling of humanity toward others and respect for oneself, an indivisible sense of the dignity of life wherever it appears." *Chun tzu*, which highlights relationships that are the opposite of the competitive, petty, and ego-centeredness. *Li* is the quality that leads to doing things correctly—in the use of language, in avoiding extremes, in the correct ordering of relationships within the family and society. *Te* is the power of moral example that attracts the willing support of the people, and it refers to the "arts of peace," specifically the power of the arts to transform human nature in ennobling ways (Smith, 1991, pp. 175–181). Taoism and Buddhism also focus on the moral nature of relationships with others and natural systems.

By way of contrast to these ancient epistemic/moral frameworks, it has only been in recent decades that Western thinkers have begun to lay the conceptual foundations for understanding the misconceptions that represent the world as material entities—both animal and human—that have their own distinct properties and that can be understood objectively and engineered to serve economic and political interests.

In Whitehead's most important and most difficult book, *Process and Reality* (1929), he challenges the idea of discrete entities or things—which range from ideas, organisms, events, material objects, facts, etc.—by claiming that actual entities are vital, transient "drops of experience, complex, and interdependent" (p. 28). That is, actual entities, contrary to the Western linguistically-driven habit of thinking of things and objects, are units of emergent processes. As he put it, "there is no going behind actual entities to find something more real" (pp. 27–28). In short, there are no self-contained "things," as everything in the human world has a history shaped by both environmental and cultural influences. Reality is best understood as ongoing relationships (units of process) that serve as creative influences on succeeding relationships.

It is the thinking of Gregory Bateson that brings into focus what is most distinctive about relationships, and to understanding a key characteristic of all ecologies. Bateson's *Steps to an Ecology of Mind* (1972) is also a difficult read, partly due to it being a collection of essays where his most important insights about relationships (ecologies) are only briefly explored and then submerged in a discussion of other non-linguistic issues. If one reads him in terms of what he has say about the interconnections between the archaic language processes we still take for granted and living systems (ecologies), the pedagogical and curricular implications begin to emerge for understanding Spretnak's observation about why the high-status systems of knowledge promoted in public schools and universities, which are largely based on print-based knowledge, misrepresent the relational world in which we live.

Key Ideas of Gregory Bateson on Language

Summaries are always dangerous, but it is possible to present Bateson's core ideas as being about how language encodes earlier misconceptions and silences that continue to marginalize awareness that relationships, and how the information communicated through these relationships, are the dominant features of all forms of existence. One of Bateson's criticisms of what he referred to as a recursive pattern of thinking in the West is the past failure to understand the individual, plant, event, data, and so forth in terms of its relationships within the ecological system of which it is a participant. The misconception that there are autonomous entities, and thus the ontological world created by this misconception, leads to studying their distinctive characteristics separate from the emergent life-altering relationships within the micro and macro ecologies that encompass all forms of life.

The following are three of Bateson's insights about language that are particularly relevant to understanding how the current educational reforms that rely more heavily upon computers reinforce the long-held cultural pattern of ignoring relationships and thus the ecology of influences that carry forward a long history of previous influences. For readers who want a deeper understanding, they should go to the chapters in *Steps to an Ecology of Mind* where Bateson speaks for himself. The section titled "Epistemology and Ontology" is the most direct discussion, although other insights are scattered throughout the book. Unlike other books on the ideas of Bateson, my book, *Perspectives on the Ideas of Gregory Bateson, Ecological Intelligence, and Educational Reforms* (2011), focuses on the connections between his insights on how the misconceptions encoded in the metaphorical nature of language perpetuate such myths as individual autonomy and the progressive nature of change, and that science and technology will enable us to survive the destruction of the environment.

Perhaps most important is how Bateson's three core ideas on language, which are largely unknown by most public school teachers, academics, and the general public, highlight how the misconceptions about a world of facts, objective knowledge, and data help us to recognize the many ways that class-room teachers and professors undermine the relational way of thinking that is essential to exercising ecological intelligence. These core ideas include:

The Map is Not the Territory

As Bateson thinks ecologically, he recognizes that everything, including words, have a history shaped by earlier cultural and environmental influences. This insight immediately brings into question how the current over-reliance upon print (whose limitations were identified earlier) undermines awareness of the ecology of language. The current meaning of words, such as woman, individualism, data, and so forth, is the outcome of an earlier process of metaphorical thinking where the analogs settled upon by think-ers in different cultural eras are carried forward and too often become the taken-for-granted basis of thinking about today's problems and possibilities. For example, the old analogs that framed the meaning of women have now, in some regions of the world, been replaced by new analogs that represent women as artists, astronauts, historians, CEOs of giant corporations, and so forth. The effort here is to reframe how to understand individuals in terms of their relationships within the larger ecologies they are dependent upon. Other cultures have already achieved a relational/ecological way of thinking about the individual, while others continue to derive their analogs from the West's consumer-oriented culture that requires the myth of individual autonomy.

The critically important issue here is how old patterns of thinking continue to misrepresent today's realities. Many of our taken-for-granted patterns of thinking continue to be based on the root metaphors (interpretative frame-works) of patriarchy (now being challenged), individualism, progress, mecha-nism, a human-centered world, economism, and now evolution, that go back hundreds of years—and in the case of patriarchy and anthropocentrism (human-centeredness), thousands of years. One of the characteristics of root metaphors is that they create supporting vocabularies that make it difficult to challenge what the root metaphor or combination of root metaphors exclude from awareness. For example, the vocabulary that supports the root metaphor of individualism, such as "freedom" and "autonomy," limits the possibility of recognizing that words have a history, and that many of the individual's taken-for-granted patterns of thinking are based on metaphors that encode the assumptions from earlier eras. In effect, the relational nature of what is mistakenly thought of as the autonomous individual needs to take account of how her/his patterns of thinking, personal identity, and even physical characteristics have been influenced by the ecologies of language, cultural identity, and genetic inheritance. The root metaphor of mechanism, which can be traced back to the thinking of seventeenth-century scientists such as Johannes Kepler, led to a vocabulary that is now used to explain organic processes, including the nature of thought itself. Other root metaphors such as evolution and progress have also led to complex vocabularies that are self-reinforcing of its deepest conceptual foundations. The excluded vocabularies limit awareness of other relationships that, as the ecological crisis deepens, are more critical to achieving a sustainable future.

If students are to learn to think relationally beyond what is required to attain immediate personal goals, which is needed for developing an ecological understanding of the world they live in, it is important for them to be introduced to Bateson's explanation of an aspect of language that has generally been ignored. That is, his explanation of what I prefer to call the linguistic colonization of the present by the past. The metaphor of "map," as he uses it, refers to the conceptual interpretative frameworks based on the vocabularies (metaphors) acquired in becoming a member of a language community. The "territory," for Bateson, refers to the current everyday world of relationships— that is, the cultural and environmental ecologies within which we live. In short, the maps (the metaphorically constructed interpretive frameworks) are generally inadequate guides for understanding and responding to current social and environmental changes. This is because the selection of analogs in the distant past, such as thinking of the environment as a source of danger and in need of being brought under human control, and then later as a natural "resource" waiting to be economically exploited, were not based on an awareness of the interdependencies between the natural and cultural ecologies. The root metaphors of mechanism and progress, which provided conceptual direction and moral legitimacy to the early stages of the scientific/industrial revolution, also limited awareness of the exhaustible nature of natural resources.

We shall later consider how students can be mentored in becoming aware of how the metaphorical nature of language illuminates or hides an awareness of what is communicated through their relationships with each other, of the traditions from the past still carried forward in their behavior and values, and of the natural systems undergoing changes that exceed the capacity of technology and science to reverse. This will be taken up when considering how current educational reforms misrepresent the ecology of language.

Double-Bind Thinking and Behaviors

Double binds were first understood by Bateson and his followers within the context of therapy situations, where the efforts to help took the form of reinforcing the very behaviors that needed to be changed—thus making the idea of progress an illusion. But the concept has more important implications in terms of understanding the double binds inherent in current widely held cultural agendas such as the globalization of the West's economic system, of digital technologies, and in the use of the English language that privileges nouns over verbs—to cite just three examples of double-bind thinking.

The linear view of progress taken for granted by the promoters of world economic growth fails to take account of environmental limits. This example of double-bind thinking leads to equating the economic exploitation of the whole biosphere we depend upon with progress. The double bind in promoting digital technologies on a global basis is that this view of progress undermines the oral traditions essential to the intergenerational renewal of the cultural commons that enable people to live more community-centered and thus interdependent lives that rely less on consumerism. In short, double-bind thinking results from relying upon the old assumptions (conceptual maps) instead of giving attention to what is being communicated in relationships that have a smaller ecological footprint.

The double bind in the process of linguistic colonization where English displaces other languages is that English nouns such as individualism, progress, intelligence, facts, environment, and so forth, reinforce a world of fixed entities that seemingly are independent of actual cultural contexts and the ecologies of emergent relationships. That is, they reproduce a static view of reality, rather than the relational/process/emergent world communicated through the use of verbs. Linguistic colonization of other cultures can be seen in how the adoption of the English vocabulary that now accompanies Western technology and consumerism within East Asian cultures, along with the printed texts of Internet technologies, are undermining their more relationally sensitive languages. As in the earlier examples, double-bind thinking fails to recognize that what is assumed to be a progressive development is in reality an ecologically destructive set of ideas and practices. Unfortunately, the language that accompanies double-bind thinking, and appears essential to a modern way of thinking, hides its own history of failure in solving fundamental social and environmental problems.

A Difference Which Makes a Difference

This phrase is part of Bateson's statement on what occurs in relationships. As it is a key to understanding both what he means by double-bind thinking and how the historically constituted conceptual maps are seldom adequate guides to understanding and responding to today's "territory," it is important to quote him in full. "A 'bit' of information," he writes, "is definable as a difference which makes a difference. Such a difference, as it travels and undergoes successive transformations in a circuit, is an elementary idea" (1972, p. 315). Bateson follows this brief statement with the example of the series of differences which make a difference such as how the axe introduces a difference in the cut-face of the tree that leads in turn to a change in the angle of the axe as it makes the next cut. The response of the Other to the difference which makes a difference can be observed in every relationship—in speaking with others, playing a game, in walking through a forest, in exploiting someone else, and so forth.

His brief statement and equally brief example are not really adequate for overcoming how we have been conditioned to think of acting on things, and to ignoring how we continually adjust our response to the difference which makes a difference in making bread, in playing a game of chess, in a conversation with others, in passing another car, in supporting the clear-cutting of an old-growth forest, in driving a car that puts on a yearly basis 8,320 pounds of carbon dioxide into the atmosphere, and in being passive as computers replace workers and further erode our privacy, and so forth. These examples of relationships encompass both cultural and natural ecologies, as well as the micro and macro scale of these interacting ecologies. And there is no escaping from them. The question is whether we can become aware of the historical linguistic influences that limit our awareness. Also, can we become aware of the ecological destructiveness of the old conceptual/cultural maps that represent individuals as rational and autonomous, and who act on the external animate and inanimate worlds? These questions should be taken seriously by everyone, but especially by teachers and people who develop curricula.

The reality is that we all adjust our thoughts and behaviors to the differences that our language and taken-for-granted interpretative frameworks enable us to recognize as we interact in the complex ecologies that are an inescapable aspect of daily life. To reiterate a key point: the emergent nature of relationships are pathways for the exchange of complex information and signs. This becomes clearer if we give attention to the multiple forms of information being communicated in changes in relationships such as in games, in a conversation, in bullying someone else, and so forth. Responding to the information Bateson refers to as "differences" is greatly influenced by the historically derived conceptual maps (metaphorical language) that influence what is recognized and what is ignored.

The over-reliance upon print and digital technologies (that is, metaphors framed by the analogs settled upon by earlier thinkers) continually reduces the emergent world to things, events, facts, and static relationships—in effect, to the world as understood in earlier eras. The metaphorical nature of language, with its historically derived analogs that frame how to interpret the world in terms of past ways of thinking, hides not only the interactive processes that are part of our living world, but also what earlier thinkers were unaware of. The culturally influenced sense of being an autonomous individual, with an inflated sense of personal agency and privilege, also leads to a reduced awareness of what is being communicated through the multiple information pathways that are part of even the most seemingly banal relationships.

Let me cite two examples of seemingly simple relationships that turn out to be complex in the different kinds of information being communicated—but mostly ignored because of cultural influences such a biases, lack of sensitivity and empathy, and the personal egos that the participants may bring to the relationship.

First, it is necessary to clarify a potential source of confusion. I have been using two metaphors, "information" and "communication," which are hangovers from the old paradigm that represented the world as distinct entities and the individual as a rational being who supposedly can provide an objective account of her/his observations of the external world. Bateson's reference to "differences which make a difference" needs to be understood as involving different messaging systems (or "information") that may range from the electrical-chemical, the genetic, differences in temperature, and so forth that influence what cells communicate to each other—and which may inhibit or promote growth. The complex physical/chemical changes in one's own bodily experience may become part of the differences (information) which make a difference in how one responds when encountering someone where tensions still exist. The connections between systems and what is communicated between them was highlighted when the 2013 Nobel Prize in Medicine was given to three researchers who discovered how hormones inside a cell, that are ferried in membrane-bound sacs known as vesicles, know how and where to deliver their genetic information so that there are no disruptions that can lead to a wide range of physical ailments. The complexity of information exchanged, for example, can be seen in how the molecular code carried in the vesicle senses calcium ions and triggers the release of brain chemicals at the right time.

The relational world of humans and animals involves even more complex semiotic/symbolic systems. In terms of cultural patterns of communication, the range of "information" generally includes non-verbal cues that send powerful messages about how the relationship is interpreted, as well as the use of words (metaphors) and silences that convey historically loaded prejudices and so forth. For example, when I tried to talk to colleagues in other academic departments about the importance of the cultural commons, the differences which made a difference for me were communicated in how quickly they averted eye contact, changed the subject, and signaled with bodily movement that they needed to go elsewhere. These differences in behavior, like all relationships, need to be understood as ecologies that were influenced by the professor's conceptual background—including influences that contributed to her/his being curious about a new way of thinking, or defensive in protecting a self-image of being a leading thinker. And these ecologies also include the ecology of language that limits or involves an expanded vocabulary necessary for understanding newly encountered ideas. The ecology of thinking within the professor's discipline, as well as the ecology of values and reward system within the department and within the discipline at the national, and even international level, all influence the professor's response to what was being communicated in the short-lived relationship.

Biosemiotics: Further Support for a Paradigm Shift

The small group of scientists who were influenced by the ideas of Bateson as well as others such as Thomas Sebeok who focused on the ecology of communication among animals, and by the growing body of research on how cells interact, are now promoting biosemiotics as a way of understanding the relational life-forming and sustaining (and destroying) processes. If the study of culture is not to be overshadowed by the continuing emphasis on the natural sciences, this new field of inquiry should be called "eco-semiotics." Referring to this new field of inquiry as eco-semiotics leads to the more inclusive understanding that all relationships, in both the natural and cultural worlds, involve some form of semiotic (information) exchange that sets in motion further exchanges.

Jesper Hoffmeyer, the Danish molecular biologist who is one of the leading thinkers in this emergent field of inquiry, reframed Bateson's statement about differences which make a difference being an elementary idea, by suggesting that the multiple forms of information communicated through differences should be understood as signs. He further shifts the focus from the traditional mechanistic way of understanding the primary characteristics of things, plants, animals, cells, and so forth, to what is occurring in their relationships. This can be seen in Hoffmeyer's observation that "the individuality of a human life cannot be justified by its uniqueness as a particular genetic combination, but must be justified by its uniqueness as a particular semiotic creature" (2008, p. 328). Thus, the individual, for example, is not to be understood only as having the capacity of being intelligent and a critical thinker, of being ego-centered, hardworking, and so forth. Instead of the personal attributes that might be identified by liberals and theologians, or by teachers, he suggests that the focus needs to shift to the biological and cultural attributes that enable participation in different semiotic systems of communication. For example, humans lack the genetic and cultural attributes that enable them to respond to the signs that enable dogs to recognize dangerous substances. Nor are the semiotic systems that Orca whales rely upon available to humans, given their differences in genetic and cultural make-up. In short, Hoffmeyer is shifting the focus from the narrow range of communication that educators and others too often associate with speaking and writing to include the whole range of life-forming processes—from the most primitive to the most complex and evolved organisms.

By introducing the idea that a more complex interspecies understanding of communication requires shifting to the more inclusive category of semiotic systems that all organisms (including humans) have the genetic and culturally mediated capacity to respond to in terms of their unique form of agency, Hoffmeyer and the others in this new field have provided a way of understanding what Bateson meant by writing that differences which make a difference represent the most basic idea or unit of information. In effect, biosemiotics (or eco-semiotics as I would prefer) is in the Whitehead and

Bateson tradition of representing reality as emergent and ongoing processes. What it adds is an evolutionary framework, and a way of understanding that the biological and cultural worlds represent different levels and forms of cognition (that is, the ability to respond to signs) at even the most elementary level.

Educational Reforms that Support the Exercise of Ecological Intelligence

The increased reliance upon the consciousness–changing characteristics of print as students spend more time reading the screens of computers, cell phones, and other digital technologies, creates a special challenge for teachers. As pointed out earlier, everyone participates in multiple ecologies—of language, cultural identities, family life, media commercialism, peer pressure, and so forth—that influence how relationships are understood—including which relationships will be ignored. In short, in taking into account the information being communicated through these relationships, everyone is exercising ecological intelligence. Awareness is most often influenced by self-interest, and what is needed to achieve immediate objectives. Some people are more aware of unjust social relationships and thus exercise what can be called a social justice oriented intelligence. And it is possible to identify a third form of ecological intelligence; one that is aware of how relationships affect the quality of life in both the cultural and natural ecologies. To reiterate another key point essential to understanding the unique challenge that today's teachers face, given the rate of climate change and the spread of poverty and unemployment that is being magnified by the digital revolution, print as a primary medium of communication is unable to represent the world as ecological systems that are emergent, relational, co-dependent, and becoming rapidly degraded. The Internet can provide vast amounts of information, but it cannot assist students in learning to interpret the short and long-term implications of what is being communicated through the multiple relationships that make up their ecological worlds. That is, the Internet relies on a sender/receiver view of language that is unable to clarify immediately that a factual statement is dependent upon metaphors that have a history. Nor can it clarify that meanings are influenced by an ecological mix of critical thinking and taken-for-granted thinking.

The starting point for helping to align how students think with the emergent and relational world within which they live is for teachers to challenge the archaic idea that they exist as autonomous beings in a world of material and unintelligent things. This can be done by introducing students to Bateson's insight that relationships are ecologies of differences that lead to reciprocal responses—in effect, a dance of information exchanges that influence subsequent behaviors. Students could be asked to observe the non-verbal patterns, as well as the changes in the use of language, that are part of every conversation and relationship. The interactive world that Bateson's phrase highlights can be seen in the differences in the behavior of a pollinating insect flying around a non-native plant. Students should be asked to give special attention to the difference which makes a difference in the behavior of the insect. That is, what are the sources of information to which the insect responds? Do the past influences include the genetic make-up of the insect as well as the plant? Why do so many people want to rid their yards of native plants? Does the absence of native plants have any relationship with the decline in the number of pollinators? How do the chemicals in the soil become critical differences which make a difference in the growth of the native flowers to which the insect responds? This may appear as leading to an inconsequential insight, but when the same question about the relationships between the toxic chemicals ingested during pregnancy and the large number of autistic infants, the importance of understanding the patterns (relationships) that connect will be recognized. Nothing exists in a totally isolated state, and the emergent patterns of interaction can be understood by giving attention to the differences which make a difference. This means giving close attention to the multiple messages being communicated in every experience, rather than being aware of only what prior print-based learning and communication establishes as being real.

Similar everyday examples, such as a sporting event, a conversation—including between people of different genders, social classes, and ethnic groups—learning from others how to plant a garden or engage in a craft, and so forth, can be used to encourage students to give close attention to the differences (information) communicated as the dance of relationships evolves.

An example that will engage the students' attention, as well as make explicit the ecology of differences that comes into play in even the most banal relationships, was suggested by Clifford Geertz (1977). In his explanation of "thick description," which is really what is being suggested here as learning to give explicit attention to the differences which make a difference (including historical and otherwise taken-for-granted patterns of influence), he suggested that his readers consider what separates an involuntary wink of the eye from the wink that is intended to send a message to another person. What then are the differences which might influence how the intended wink is understood and responded to, or behaviors that follow from a series of misunderstandings? What are the behavioral and other changes occurring in the local context? How does memory influence how the relationships prompted by the wink will evolve—and even be misunderstood? How do gender and social status differences become part of the message exchange?

Another common everyday relationship that involves multiple messages that can lead to misunderstandings, depending on the largely taken-for-granted cultural assumptions the participants bring to the relationship, is the way that people engage in different forms of physical contact. The growing tendency toward engaging in physical embraces is an example of ecologically complex messages—that is, differences that should have made a difference where what is ignored could become a new set of differences that become part of a new succession of differences that undergo "transformations in a circuit" (to get back to Bateson's wording). Having students observe how and when people embrace each other, as well as the non-verbal patterns of communication that follow, provides yet another example of the complex range of transformation in the differences which make a difference. It will also provide a good example of what Spretnak and others are saying about living in a world of relationships—an awareness that may lead to reducing the mindless behaviors that set off a string of consequences that go unnoticed when the complexities of relationships are ignored.

Part of understanding how so much of the conceptual world in the West misrepresents the emergent and relational world of everyday existence can be addressed if teachers encourage students to understand the fundamental differences between face-to-face communication and oral cultural storage, and how a static view of the world emerges from print-based storage and communication. In helping students to understand the differences it is important to emphasize the dangers of either/or thinking, which might lead them to conclude that print-based knowledge or face-to-face communication should be abandoned. The importance of each depends upon the context, and ultimately on which contributes to community self-reliance and an ecologically sustainable future. It would also be useful for students to understand why print-based cultural storage and thinking is inherently ethnocentric. This might enable older students to recognize why so much of Western philosophy and social theory has had a colonizing influence on other cultures—and why the digital revolution is having the same impact.

The point made earlier about how language carries forward the misconceptions and silences from the past also has implications for teachers who realize that computer-based education, both at the instructional and testing level, indoctrinates students to accept the mindset promoted by the digital revolution—which has an anti-democratic and pro-corporate agenda that is ignored because of the many genuine contributions of digital technologies. If the historically encoded vocabulary the students acquire limits their awareness of what is being communicated through their relationships within the larger cultural ecology in which they live, then it should be obvious what the teachers' responsibility should be. They should help students recognize the metaphorical nature of the language/thought connection, including how the current meanings of words were often framed by the analogs settled upon in an earlier cultural era. This should include helping students understand how to reframe the meaning of words such as "traditions," "intelligence," "data," "markets", "wealth," "individualism," and so forth, by selecting new analogs that are ecologically and culturally informed.

That is, it is important for students to be able to recognize the world more as it is rather than to filter it through the interpretative frameworks influenced by a language that encodes the prejudices and silences of earlier generations. For example, the word "tradition" is a metaphor that still carries forward the misconceptions of Enlightenment thinkers who were unaware of the importance of the cultural commons of their day (which will become even more important today as computers replace the need for workers). They were also unable to anticipate how encoding the word "tradition" with their optimism about rational thought and technological change would lead to today's loss of privacy and historical memory of how to live less money-dependent lives.

The exercise of ecological intelligence requires being aware of what is being communicated through the relational information pathways (or through the "differences which make a difference," to quote Bateson again) and recognizing when the information is a sign of a destructive relationship. As universities continue to ignore engaging students in a deep understanding of the cultural amplification and reduction characteristics of technologies, which include computers, print and visual media, it is unlikely that we will be able to escape the inherited taken-for-granted conceptual patterns that each generation passed forward and disguised as the latest expression of progressive thinking. Unfortunately, these patterns of thinking led to the first industrial revolution, which is likely to be overlooked now that the digital revolution is contributing to the cultural amnesia that is eliminating memory of relationships and patterns of mutual support that were less destructive of cultural and natural ecologies.

References

Bateson, G. 1972. *Steps to an Ecology of Mind.* New York. Ballantine.

Bowers, C. 2011. *Perspectives on the Ideas of Gregory Bateson, Ecological Intelligence, and Educational Reforms.* Eugene, OR: Eco-Justice Press.

Geertz, C. 1977. *The Interpretation of Cultures.* New York: Basic Books.

Hoffmeyer, J. 2008. *A Legacy for Living Systems. Gregory Bateson as Precursor to Biosemiotics.* Dordrecht: Springer.

Nisbett, R. 2003. *The Geography of Thought: How Asians and Westerners Think Differently . . . and Why.* New York: Free Press.

Smith, H. 1991. *The World's Religions.* New York: HarperCollins.

Spretnak, C. 2011. *Relational Reality: New Discoveries of Interrelatedness that are Transforming the World.* Topsham, ME: Green Horizon Books.

Whitehead, A. 1929/2010. *Process and Reality.* New York: Simon and Schuster.

4 Language Issues that Should be the Central Focus in Teacher Education and Curriculum Studies

Regardless of the specialized area of interest—from becoming a reading specialist, English or social studies teacher, an environmental educator, a math and physics teacher—all teachers will be dependent upon the cultural processes seldom addressed directly in teacher education and graduate educational studies programs. For that matter, the education of academics in other disciplines, with only a few exceptions, perpetuate the same silences. Largely ignored are the different languaging processes that reproduce a culture's taken for granted ways of knowing, values, and understanding of human and nature relationships. There is, in short, no area of the curriculum that does not rely upon the languaging systems of the culture, or that of other cultures. Yet learning about these processes, what they hide, what they distort, when they become sources of empowerment, and, more importantly, when they contribute to cultural changes that may reduce the culture's adverse impact on natural systems, are seldom part of the classroom teacher and professor's professional studies.

When as a member of the faculty at the University of Oregon I tried to introduce this major shortcoming in the teacher education program I was told that I should understand that since everybody uses language to communicate their ideas there was nothing further that needed to be studied. They were thinking of language as a conduit in a sender/receiver process of sending ideas, data, and information to others—which is the view of language reinforced in the use of textbooks, in computer-mediated communication and software programs, in lectures, in power-point presentations, and in most daily conversations. Unfortunately, this view of language is incorrect, and is a major reason why so little attention is given in educational settings to how the deep ecologically problematic cultural assumptions formed in the distant past continue to be passed from generation to generation. Given the long progressive tradition of educators urging students to construct their own knowledge, which a large segment of the public is now doing in ways that threaten the historical foundations of our democracy, there is even less awareness of the range of language issues that teachers and professors should be addressing beyond those relating to gender, racial, and homophobic discrimination.

In a recent course I taught at the University of Oregon, I avoided introducing students to a survey of readings that address curriculum issues, as I knew they would encounter these surveys in other courses. I also avoided the mistake made in a newly established computers in education course where students were asked to come up with their own reading list. By not knowing the issues relating to the cultural non-neutrality of computer-mediated learning, or the questions that should be asked about the ideology encoded in software programs, the students presented 20 minute summaries that reproduced most of the silences that exist in the current educational discourse about the advantages of relying upon computers in the classroom. They were pleased to be in control of their own learning, but were totally unaware of the questions and conceptual frameworks necessary for understanding what they need to bring to the attention of students about the cultural non-neutrality of computers and digital communication in general.

My many years of challenging what I earlier referred to as the culture of denial has led to the conviction that teacher education courses, as well as those I taught in the Honors College and the Center for Environmental Studies, should introduce students to what they do not know: that is, the cultural silences perpetuated in classrooms and through the media. The course described below thus avoids introducing students to past theorists who wrote about different approaches to educational reforms that were relevant in earlier days but totally irrelevant in terms of addressing today's issues. It also avoids what too many students now expect: to control what they learn and to not be too intellectually challenged. Given the narcissistic mind-set of many of their professors, these expectations seem on the surface to be praiseworthy yet naïve in a world facing environmental and cultural challenges not encountered before. The following is how the students were introduced to an in-depth understanding of the languaging processes that are unavoidable in making pedagogical and curricular decisions. It also presents the core concepts that students in professional education courses should encounter before taking other courses—especially survey courses. The justification for this claim is that the concepts explored in depth in the course lead to recognizing the 20th century legacy of misconceptions and silences that will be encountered in the specialized courses in teacher education and in other disciplines across the university.

EDST 610 Curriculum Reform for a Sustainable Future
Overview of Course

In taking Albert Einstein's observation seriously that the same mind-set that created the problem cannot be relied upon to fix it, this course will have four main foci. **First,** *it will reframe the current approaches to thinking about curriculum reform in ways that take account of the cultural/linguistic patterns of thinking and relationships that contribute to a smaller ecological footprint, and to lifestyle changes that address the growing unemployment due to the further automation of the workplace. Special* *attention will be given to what teachers need to understand about how the language in the curriculum and in classroom discussions often reproduces the misconceptions of an earlier era when environmental limits were not understood. How to help students recognize when it is important to reframe the meaning of words in ways that are culturally and ecologically informed will also be given attention.* **Second,** *attention will be given to how curriculum reform can help students recognize the connections between a consumer-dependent lifestyle and the deepening ecological crises. The nature and ecological importance of the local cultural commons (the intergenerational knowledge, skills, and mentoring relationships that are less dependent upon consumerism) will also be considered, as well as the teacher's role in helping students become more aware of the differences in their personal development and the ecological impact as they move between the relationships and activities within the local cultural commons and settings where they are consumers.* **Third,** *attention will be given to what students need to understand about how computer-mediated learning contributes to a smaller ecological footprint within certain contexts as well as how it undermines the local cultural commons. How to incorporate into the curriculum an understanding of the cultural transforming characteristics of computers will also be addressed.* **Fourth,** *attention will be given to how to understand the nature of ecological intelligence, and how it differs from the myth of individual intelligence that is reinforced in most classrooms as well as in print-based forms of communication. Understanding these core issues will also be sources of empowerment in other work settings.*

This approach to introducing future teachers to the core issues in their profession avoids treating the language issues in the abstract or only in terms of social justice issues. The main focus was on the wide range of issues that are ignored because of the conduit view of language that is dominant in so many areas of discourse (both written and spoken). The discussion also focused on curricular approaches that would enable their students to recognize how language shapes their own experiences in ways that challenge the long-held assumption that autonomous thinking individuals use language to communicate their ideas to others—and that there is such a thing as objective knowledge and facts. These are among the most important misconceptions that classroom teachers must know how to address in ways that take into account ethnic differences as well as the even greater challenge of enabling students to recognize the role that language plays in deepening the ecological crisis. Bateson's insight that understanding relationships rather than focusing on individual entities was used as the basis for putting the teachers' professional knowledge on an ecological footing.

That is, the four major conceptual categories mentioned in the description of the course were presented as ecologies—the ecology of languaging processes, the ecology of the tension between the cultural commons and a consumer-dependent lifestyle, the ecology of computer-mediated learning and abstract thinking, and the ecology of moving to different levels in exercising ecological intelligence. One of the implications of reframing what is being suggested here as the conceptual areas that should be part of the basis of the teacher's professional knowledge is a characteristic of all cultural and natural ecologies: namely, that relationships within micro and macro ecological systems have a history, are sustained though complex patterns of message exchanges, and that both their history and current patterns of interaction and interdependence have implications for their future prospects. These relationships are essential for thinking about the ecology of language, of the cultural commons, of the print/computer/Internet patterns of storage and communication, and of the process of exercising ecological intelligence.

The discussions on curriculum reform, particularly those that focused on the pedagogical and curricular implications, stressed that helping students to think ecologically requires that they take into account the history of ideas, words, events, technological developments and so forth—as well as how they affect current patterns of interaction and mutual support, as well as their implications for the future. These simple and common sense guidelines for thinking, as discussed in the class, brought into question the current emphasis (largely driven by testing and most teachers' past socialization) that there are objective facts, that events and ideas can be understood in terms of cause and effect, and that abstract ideas and accounts of events have universal relevance. The following represent the key concepts that were discussed in the course, with a major focus being on how the concepts can be introduced to public school and university students. It was emphasized that the concepts should be reframed as questions to be investigated in term of the student's own cultural contexts. In addition to the importance of the concepts, question-directed inquiry encourages students to give close attention to cultural patterns they might otherwise take for granted. The questions, when raised in the classroom, are really a matter of naming what previously was un-named and thus not made explicit. And when made explicit, the connections between otherwise taken for granted cultural patterns can then be examined in terms of their consequences on other patterns. Understanding the patterns that connect, as Bateson reminds us, is learning to think ecologically. This process then needs to be extended to examining how the student's taken for granted cultural patterns of thought and behavior affects the ecology of relationships within the natural systems. Indeed, this might be easier for younger students to do than adults whose ideas and patterns of behavior have become habituated.

Ecology of Languaging Processes

1   Words have a history (which is a basic insight that leads to classroom discussions of the following curricular possibilities).
2   Most words are metaphors that encode the analogs settled upon in earlier times and thus carry forward the insights, misconceptions, and silences of the earlier state of cultural awareness. (Student conducted interviews will yield short term perspectives on how the meaning of words have changed, and the Online Etymological Dictionary will provide a

long-term perspective on how the meaning of words have changed over time—and been influenced by other linguistic traditions).

3   When born into the metaphorical language of a language community, the initial process of thinking is influenced by the historically constituted meanings that others took for granted. That is, acquiring the language of one's community also involves being dependent upon ways of thinking about issues and problems that were unknown in earlier times.

4   This disconnect between past ways of thinking and current ecological realities is the basis of double bind thinking.

5   The taken for granted ways in which people rely upon past ways of thinking, even while thinking that their ideas and values are individually determined, are a dominant characteristic of the curriculum—whether spoken or read.

6   Most curricula and most patterns of verbal communication reinforce a conduit view of language where supposedly objective ideas and data, or an individual's own ideas, are passed to others.

7   The conduit view of language, which is reinforced in print and thus in computer-mediated cultural storage and communication, hides the basic reality that words have a history—as well as how they carry forward earlier culturally specific ways of thinking, misconceptions, prejudices, and silences.

8   The conduit view of languaging processes hides how words, as metaphors, encode earlier and culturally specific analogs that contribute to the linguistic colonization of the present by the past, and to the colonization of other cultures.

9   Words can be given new meanings when the choice of analogs is informed by other cultural ways of knowing and a knowledge of current environmental changes.

10   Learning about the history of words, as well as considering whether the analogs derived from earlier ways of thinking are adequate for understanding and responding to current cultural and environmental issues, will enable a diverse student group to understand both the linguistic colonization of the present by the past as well as the linguistic colonization of students learning to speak English.

Ecology of the Cultural Commons

1   The cultural commons involve the daily practices that are based on intergenerational knowledge, skills, and patterns of mutual support that rely less on a money economy.

2   The cultural commons exist in every community and the activities range from food preparation and sharing, healing practices, patterns of mutual support, ceremonies, games that depend upon the rules handed

down from the past, creative arts and craft skills that rely upon mentoring, a heritage of knowledge of place, civil liberties, knowledge about the care for animals, how to build a dwelling, to live by values and skills that have a small impact on the land, and so forth.

3 Cultural commons activities bring people together in interdependent relationships, and lead to discovery of personal talents and skills.

4 The cultural commons, that is the intergenerational knowledge that enabled pre-industrial people to survive, grow in population, and to expand their knowledge and patterns of self-governance, began with the first humans wandering the savannas of what we now call Africa.

5 The cultural commons involve a more complex economy of mutual exchange, barter, and volunteerism, and thus provides for alternative community-centered lifestyles in an era of increasing automation driven unemployment and economic uncertainties.

6 Most cultural commons activities have a smaller toxic and carbon impact on natural systems.

7 The cultural commons are as diverse as the world's cultures and bioregions.

8 Practices within the cultural commons more often involve local decision making, and a sense of making decisions that strengthen community—rather than decisions based on the pursuit of self-interest, competition, and more profits.

9 Public schools and universities tend to marginalize awareness of the local cultural commons by emphasizing abstract thinking (that is, print-based knowledge) as well as the values and assumptions that underlie an individualistic, progressive, and consumer-oriented society.

10 Many aspects of different cultural commons, especially their traditions of narratives and patterns of mutual support, carry forward prejudices and patterns of discrimination—thus the cultural commons should not be romanticized.

11 The world's diversity of cultural commons are being undermined (enclosed) by technological and market forces that are working to integrate them into the consumer culture.

12 Computer driven automation is also turning more aspects of the cultural commons into services and products that require dependence upon the money economy that benefits the already wealthy at the further expense of the poor.

13 The teacher's role as a mediator is to help students articulate the differences in the development of personal talents, ecological footprint, and patterns of mutual support as they move between their cultural commons and market-based experiences.

14 One of the goals of the teacher's mediating role is to help students become more aware of the community's traditions of self-sufficiency and mutual support, and thus to be able to recognize when technologies and market forces threaten to overturn these traditions.

15 In helping students articulate these differences, as well as recognize aspects of the scientific and industrial culture that have made positive contributions to humankind, the teacher is addressing a fundamental problem in our increasingly complex democracy: namely, the ability on the part of the student to acquire the linguistic and conceptual basis for exercising communicative competence in the political process.

Ecology of Print-Based Storage and Computer-Mediated Learning

1 Print-based cultural storage and thinking have radically different affects on consciousness and social relations—which are not understood by most public school teachers and university professors.

2 Print-based cultural storage and thinking have the following characteristics that are, in turn, influenced by differences in cultural ways of knowing:

   a Print can only provide a surface understanding of ideas, events, and processes.

   b Print is unable to represent the deep cultural and natural ecologies of information that underlie the origin and current influence of ideas, events, and processes that are inadequately represented by the word "context."

   c What is committed to print becomes immediately outdated in a world of cultural and natural ecological systems that involve both historical continuities and constant change.

   d Print reinforces abstract thinking and thus the tendency to treat abstract thinking as representing universals that no longer take account of different cultural and natural contexts.

   e Printed texts, whether in a book or on a computer screen, reinforce the conduit view of language that, in turn, hides the metaphorical and thus historical forces that continue to frame the meaning of many words.

   f One of the aspects of print that has a powerful influence on consciousness, and thus on social policies, is that print allows people who would otherwise be constrained by the misconceptions of their community to communicate their ideas to a broader and even a future audience.

3 Print too often is interpreted by the reader and even the writer as representing an objective account of reality.

4 Print reinforces key characteristics of modern western culture: the validity of the individual's perspective and critical analysis, sight as the source of knowledge, objective knowledge that can be universalized.

5 Printed accounts marginalize the information acquired through the senses other than sight, which print privileges.

6   Print has served as a key part of the process of colonization, in terms of maps that designate political boundaries without awareness of cultural differences of the groups within these boundaries, and the written treaties that do not take account of the cultural traditions of the groups constrained by the written treaties.

7   Literacy has served as justification for colonizing oral cultures that were seen as uncivilized and thus in need of being brought into the modern world which is largely dominated by abstract thinking and political ideologies.

8   Computer-mediated learning and thinking, as well as reliance upon other forms of Internet-based communication, reinforce the sense of being an autonomous individual and in control of where one wants to go in terms of cyberspace and thus the future.

9   Unlike oral cultures where elder wisdom and narratives carry forward the moral values and, in many instances, the knowledge of how to live within the limits and possibilities of the local bioregion, computer and Internet based thinking marginalizes awareness that everything in the cultural and natural ecologies has a history—with implications for the prospects of an ecologically sustainable future.

10 One of the dominant influences of computer and Internet mediated thinking and communication upon consciousness is that cultural amnesia is becoming more widespread—which can also be understood as the loss of long-term memory.

11 Oral patterns of storage, thinking, and communication more often rely upon all the senses, and not just sight, and are more likely to avoid many of the limiting characteristics of print. Other implications include the following:

   a   Relying upon all the senses provides more access to the information being exchanged in the local cultural and natural ecologies.
   b   Oral communication increases awareness of ongoing relationships.
   c   Oral communication also relies more on active memory of the other participants in the relationship.
   d   Oral cultures are more community-centered which can lead to practices of mutual support, as well as discriminatory practice of members who are viewed as deviating from the shared moral norms. Tolerance of differences varies from culture to culture.
   e   Oral cultures vary in terms of their restrictions and allowances for participatory decision making, but they also have been more successful in the past of socializing the new generations to the moral norms governing human/nature relationships. This is now disappearing with the spread of literacy and the accompanying sense of individual autonomy.
   f   Oral cultures are coming under increasing pressure to adopt the print-based form of consciousness, which also includes adopting

the western technologies that are alienating the youth of these cultures from the intergenerational traditions that underlie the cultural and natural commons.

Ecological Intelligence

1   Scientists reduced the ancient Greek concept of "oikos" to the study and management of the natural environment, which subsequently became known as "ecology."

2   As Gregory Bateson points out, all systems, from the micro to the macro natural systems—and including cultural patterns of information exchange—are ecologies.

3   Just as there are ecologies of weeds, there are cultural ecologies based on misconceptions carried forward in the ecology of language.

4   One of these misconceptions that has led to important developments in the areas of civil liberties, but also to destructive developments in how the West has exploited the environment, is the idea that intelligence is an attribute of the autonomous individual.

5   Thinking of an individually-centered exercise of intelligence does not take account of how thinking, values, and behaviors are influenced by the metaphorical and thus historically influenced language acquired in becoming a member of the language community.

6   The key to understanding that all relationships within cultural and natural systems are a central feature of all living ecologies is the information that is in constant circulation within the ecology's subsystems—which may be at level of the genetic/electro-chemical exchanges that sustain biological systems and in the conversations between speakers. The responses that sustain life processes are triggered by what Bateson refers to as the "difference which makes a difference" in the response of the Other.

7   Human with human and human with nature relationships always, to varying degrees, take account of the differences communicated by the Other—which may be a change in the weather, the gesture or tone of voice of the Other, the off-key performance of other musicians, the use of words that have no relationship with other experienced realities, the introduction of a chemical that has a toxic effect on a biological system, and so forth.

8   Living in a world of ongoing and evolving relationships, where we are constantly aware and responding to differences (people walking on the wrong side of the street, the body language of the Other, weather too extreme for the season, the absence of animal sounds, and so forth) means that everybody exercises in varying degrees ecological intelligence.

9   The exercise of ecological intelligence is not based on abstract thinking, but on the recognition that people are in varying degrees aware and

respond to what is being communicated in their relationships within the cultural and natural ecological systems.

10 Playing a game, driving a car, cooking a meal, working with clay, interacting with students engaged in computer-mediated learning, and so forth, are the everyday examples of exercising ecological intelligence. There is always a question of whether awareness of relationships extends to those that are genuinely life-sustaining of the larger systems we are all dependent upon.

11 The misconceptions and silences encoded in the language influence which differences which make a difference will be recognized and which ignored, and how the differences are interpreted. There is constant communication about the impact of the carbon footprint of Americans on natural systems—including the growing acidification of the world's oceans and the melting of the Artic ice—yet most Americans, while responding to other differences to which they have learned to observe and respond, continue to purchase vehicles that add to the carbon foot print.

12 The exercise of ecological intelligence should be understood as involving at least three different levels, though in some cultures the deep assumptions make the exercise of stage-three ecological intelligence a common feature of everyday life. What is noteworthy is that many of these cultures did not rely on western technologies that promote abstract thinking, the myth of individual autonomy and technologically driven progress that are now undermining these complex traditions of sustainable thinking and behavior.

13 The taken for granted cultural assumptions underlying the individual/consumer-dependent lifestyle promoted by the industrial culture leads to an individually-centered exercise of ecological intelligence. The information circulating through the relations is selectively perceived, and understood as part of strategy for achieving one's personal goals.

14 Stage-two ecological intelligence involves giving attention to the difference that make a difference—such as being aware of inadequate diets among children, the way automation is displacing the need for workers, the efforts to suppress the ability of the poor and marginalized to vote in an election, and so forth. Taking this information into account leads to addressing social injustice issues.

15 Stage-three ecological intelligence involves awareness of social justice issues but also a concern with how to conserve as well as how to initiate changes that contribute to a sustainable future—for both the cultural and natural systems. Awareness that hyperconsumerism, and the language that provided conceptual direction to the Industrial Revolution that is now in its digital phase of globalization, are ecologically unsustainable would come naturally to a person who exercises stage-three ecological intelligence.

16 People who are more oriented to exercising sustainable ecological intelligence may at times operate at the level of individually-centered ecological intelligence, and even at the social justice level. In some cultures, stage two and three are a taken for granted ways of responding to the differences which make a difference in their lived ecologies.

17 A major obstacle to students becoming aware of the different stages of ecological intelligence is the modernizing ideology promoted in teacher education programs that go under different labels but share many of the same deep cultural assumptions of the market liberals who are promoting western assumptions and values as the basis of a new global economic order. These include critical pedagogy, transformative learning, place based education (which does not question the myth of individual intelligence), eco-pedagogy, and computer-mediated learning that reinforces abstract thinking and the idea that students should construct their own knowledge.

Implications that go beyond Teacher Education and Curriculum Classes

Due to the time constraints of the summer class, as well as the pressure students were under from other classes, the exchanges indicated that some students were not doing all the readings, or were treating them too superficially. Nevertheless, most of the key ideas that brought a depth of understanding of the overall conceptual framework were briefly discussed, with many of the more important ones being examined in terms of how they lead to a more explicit awareness of how the curriculum, both spoken and written, reinforces the ecologically destructive patterns of thinking and values. There were also extended discussions of how key ideas lead to introducing curriculum changes at different levels in the educational process. A point that was stressed is that the curriculum reforms should enable students to understand that words have a history and encode earlier ways of thinking, that the cultural commons represent alternatives to a consumer-dependent lifestyle, that print-based storage (including Internet-based thinking and communication) undermines the exercise of ecological intelligence.

What is distinctive about making these language issues a central part of the teacher's professional knowledge, and which should be at the center of any university reform effort, is that the explanatory power of the key concepts listed above is not acquired by reading textbooks or downloading explanations from Google. Rather, the concepts help make explicit how the languaging processes in the curriculum reinforce the old patterns of thinking and limited awareness, or enable students to recognize patterns that before were hidden behind the fog of their taken for granted experience. That is, the emphasis was on how the concepts could be used to frame which aspects of the students' otherwise taken for granted culture could be made explicit, described, and then examined in terms of its impact on their own lives, the effect on the well-being of the community and on natural systems. One of the key ideas in the above list is that reliance upon print-based descriptions leads to abstract thinking, and thus to a surface knowledge that does not take account of differences which make a difference within the local environmental and cultural ecologies.

The point was stressed over and over again that after introducing students to a brief introductory explanation that frames what the concept brings to awareness the focus should then shift to doing a deep-ethnography (similar to what Geertz referred to as a "thick description") of the patterns of behavior, thinking, and valuing that collectively constitute the living cultural and natural ecologies the students are embedded in. The teacher's role is that of a mediator who asks the questions, provides students the conceptual space to give words to what otherwise is part of their unrecognized and unarticulated taken for granted world, and at times reframes the issues in ways that help make explicit other processes and possible consequences the students may not have considered.

For example, asking students from different ethnic backgrounds to identify the analogs within their own culture that frame the meaning of words such as "tradition," "education," "sacred," "intelligence," "poverty," and so forth, and to compare these analogs with what most speakers of English take for granted can easily lead to a examination of how print serves to hide the cultural history of words—of both the dominant and ethnic cultures. After a careful and in-depth examination of the student's cultural commons, as well as what is shared across cultural groups, the teacher can ask students to do an in-depth ethnography of how the various Internet technologies that rely upon print and other abstract forms of representation affects awareness of the intergenerational communication that carries forward the cultural commons. The question that needs to be asked is how computer-mediated learning and participation in the many social networking systems contribute to undermining the cultural commons. This, in turn, will lead back to making explicit aspects of the students' culturally (that is, ethnically) mediated experience as well as to considering patterns they have observed within the larger society—to which they have not given serious thought.

The importance of relying upon the list of key concepts about the languaging processes that characterize all cultural ecologies is that they point to the need for students to learn about their own context-specific cultural patterns that affect their communities and the natural systems that are now being degraded. This is radically different from forcing students to acquire the abstract information that is framed by writers who are mostly unaware of the cultural assumptions that influence the interpretations they too often misrepresent as objective facts and information. The challenge is whether the current and future generations of students entering the teaching profession can recognize the misconceptions of their professors who are still under the influence of the conceptual orthodoxies of the late 20th century when there was little awareness of the cultural roots of the ecological crisis, and the ways in which print-based knowledge reproduce a surface knowledge of local contexts—as well as the silences in the thinking of the experts who produced it.

References

Bateson, G. 1972, *Steps to an Ecology of Mind.* New York: Ballantine Books.
_____. 1980. *Mind and Nature: A Necessary Unity.* New York. Ballantine Books.

5   Why the George Lakoff and Mark Johnson Theory of Metaphorical Thinking Fails to Address the Linguistic Issues Related to the Ecological Crisis

The alternative to thinking of language as a conduit in a sender/receiver process of communication, in which words supposedly accurately represent real things, events, and relationships, is to recognize that words have a history and that their current meanings are framed, for the most part, by analogs chosen in the past. As language, both spoken and written, is an inescapable aspect of education, especially as it exists in the classroom, it is important that professors and classroom teachers understand how language reproduces in the present earlier forms of intelligence—particularly those forms of intelligence that did not take account of environmental limits or other cultural ways of knowing. The growing awareness of the metaphorical nature of the language/thought/ cultural connections has been influenced in recent years by the writings of George Lakoff and Mark Johnson. Thus, it is important to clarify why their theory of embodied experiences as the source of metaphorical thinking is fundamentally incorrect. Examining the sources of their misconceptions will also bring into focus the connections between relying upon the analogs settled upon in the past that continue to frame the current meaning of words and the need to reframe the meaning of words by identifying ecologically and culturally informed analogs. The challenge of identifying analogs that reflect current social justice thinking (which occurred in reframing the meaning of metaphors such as "woman") as well as a more complex understanding of ecologies, is one that all professors, regardless of their discipline, need to address.

George Lakoff and Mark Johnson set out to radically change one of the dominant traditions of western philosophy, which is the tradition of abstract theory that stretches from the ancient Greeks down through the writings of the contemporary analytic philosophers. In place of de-contextualized and thus culturally uninformed theories about the nature of reality, mind, language, and individualism, they argue that the task of the philosopher is to clarify how the metaphorical basis of language, and thus systems of knowing, originates in the embodied experience of individuals. Their agenda is summed up in the title of their major book, *Philosophy in the Flesh: The Embodied Mind and Its Challenge to Western Thought.* (1999) While they make a cogent case against the many ways in which western philosophers have framed the process of reasoning, thus achieving little more for humanity than giving legitimacy to their own theoretical edifices, the Lakoff and Johnson argument that metaphorical reasoning originates in the individual's "sensorimotor experiences so regularly they become neurally linked" (1999, p. 555) represents an equally extreme and problematic position. In place of the rational process represented by most western philosophers as free of both cultural and embodied experiences, Lakoff and Johnson argue that the starting point of philosophy, science, and knowledge generally begins with the individual's perceptual and motor systems—that is, her/his embodied experience. In his recent book, *The Meaning of the Body: Aesthetics of Human Understanding* (2007), Johnson introduces the phrase "organism-environment coupling" to reaffirm the earlier position he shares with Lakoff that knowledge of the environment is limited to the embodied experience of the individual. As will be explained later, this precludes learning about the environment from scientific studies and from the observations and insights of others.

The word culture occasionally appears in their joint writings on how language is framed by embodied experience, but its complex nature and diversity is not explored in any depth. One of the results is that the implications of a comparative study of different cultural epistemologies has for bringing into the question the western notion of the autonomous (that is, the supposedly culturally uninfluenced) individual they take for granted—and upon which their entire theory rests—are not considered. A word they do not mention is "ecology." This omission leads to their failure to acknowledge that today the ecological crisis should frame any discussion of metaphorical thinking. What cannot be explained by their theory is why they share this oversight with key western philosophers who also ignored how the environments of their era were being degraded. Supposedly, the individual whose sensorimotor experiences and habituated neural connections become the basis for framing the meaning of words (metaphors), and thus for how relationships are understood, is unaffected by the global changes in the natural environment.

This is not simply an oversight which has few if any serious implications. It becomes of paramount importance when it is recognized that the extrapolation of the word "ecology," which is the modernized version of the early Greek word "oikos," always situates the individual as a participant within the context of cultural and natural systems. It is only when the individual is treated as an abstraction that these ecological relationships are ignored. In effect, the individual's embodied participation in this larger ecology of relationships includes other people, the semiotic systems of the culturally constructed world, and the ongoing message exchanges (which Gregory Bateson refers to as the "difference which makes a difference," 1972, p. 318) that sustain the complex and interdependent living systems we refer to as the natural environment.

In order to understand the long-term problem of locating, as Lakoff and Johnson put it, "our conceptual system" in the individual's "perceptual and motor systems" (1999, p. 555), it is necessary to summarize the changes that the Earth's natural systems are undergoing. Their extreme reductionist understanding of the origins of knowledge leads to a radical difference between what individuals would learn from their embodied encounters with their local environments and what scientists are now reporting. For example, scientists studying the impact of global warming have documented that the Greenland ice cap is melting at an accelerating rate, with one glacier moving to the sea at a rate of 2 meters an hour on a 3 mile front and at a depth of 1500 meters. The melting of the Arctic sea ice and the glaciers in the Antarctic is also accelerating at a rate totally beyond what scientists thought possible. And the glaciers in the Himalayas and the Tibet-Qinghai Plateau that feed the major rivers in India and China are disappearing at a rate of 7 percent a year, with glaciers in other parts of the world disappearing at a similar rate. (Brown, 2008) The basic assumptions of Lakoff and Johnson (listed below) limit the individual's conceptual understanding of the environment to her/his embodied experience. If the individual's embodied experience is not in an environment that is undergoing the changes cited above, she/he would be unable to grasp the world-wide consequences of global warming.

Since philosophers have a long history of ignoring how cultural belief systems impact the life sustaining capacity of local ecosystems, the question that is likely to come up is: what relevance does an overview of the ecological crises have for assessing what is problematic about the Lakoff/Johnson theory of the embodied origins of our guiding metaphors? What is being overlooked by the scientists and engineers who are trying to develop more sustainable and less carbon-producing technologies, and by the general public that has accepted that new technologies are the solution to the ecological crises, is that we need to change the metaphorical language that gave conceptual direction and moral legitimacy to the industrial/consumer-oriented culture that has become a major contributor to overshooting the sustaining capacity of natural systems. Lakoff and Johnson got it right when they argued in *Metaphors We Live By* (1980) that all thought is based on metaphors. When they made the turn toward locating the source of metaphors in the embodied experience of the individual, which was motivated by their concern with the hegemony of abstract theory and language usage by mainstream western philosophers, they lost sight of the more obvious and now ecologically important characteristic of language. That is, they ignored that words, as metaphors, have a history. In effect, they failed to recognize that the industrial/consumer-oriented culture that is now being globalized, and that is overshooting the sustaining capacity of the natural systems, is based on the metaphorical thinking of earlier thinkers who were unaware of environmental limits.

Before explaining how much of today's ecologically problematic thinking is based on what Gregory Bateson refers to as "double bind thinking," it is necessary to reproduce here the six basic assumptions that the Lakoff and Johnson theory is based upon. In order to avoid any misrepresentation of their assumptions, the assumptions shall be presented as they appear in *Philosophy in the Flesh*.

Embodied Reason

- Embodied Concepts: Our conceptual system is grounded in, neurally makes use of, and is critically shaped by our perceptual and motor systems.
- Conceptualization Only Through the Body: We can only form concepts through the body. Therefore, every understanding that we can have of the world, ourselves, and others can only be formed in terms of concepts shaped by our bodies.
- Basic-Level Concepts: These concepts use our perceptual, imaging, and motor systems to characterize our optimal functioning in everyday life. This is the level at which we are maximally in touch with the reality of our environment.
- Embodied Reason: Major forms of rational inference are instances of sensorimotor inferences.
- Embodied Truth and Knowledge: Because our ideas are framed in terms of our unconscious embodied conceptual systems, truth and knowledge depend upon embodied understanding.
- Embodied Mind: Because concepts and reason both derive from, and make use of, the sensorimotor system, the mind is not separate from or independent of the body. Therefore, classical faculty psychology is incorrect. (1999, p. 555)

There can be no doubt that many of our metaphors have their origins in bodily experiences, as Lakoff and Johnson point out. Concepts such as up and down, back and forward, full and empty, and even the old British systems of measurement of inch, foot, yard, and mile can be traced back to bodily experiences. Also, their discussion of how different experiences provide generative frameworks (schemas) for understanding an activity, behavior, and policy, in which the already familiar becomes the model for understanding something new, needs to be taken seriously. It needs to be pointed out, however, that their insight is only partially correct. Some of our concepts do have an embodied origin yet even these conceptual schemas will differ from culture to culture, depending upon the culture's mythopoetic narratives and/or cosmology. For example, while Lakoff and Johnson would attribute the concept that underlies the use of the personal pronoun "I" to the embodied experience of an individual, they overlook that this is a culturally constructed identity—one that can be traced to the writings of post-medieval philosophers and political theorists. Instead of the "I want" and "I think" habituated pattern of thinking so prevalent in the West, there are profoundly different ways of understanding self which vary from culture to culture. Among the traditional Maori, for example, when a guest enters into the marea (the communal gathering place), she gives her name and then her lineage—followed by an explanation of her relationship to the family or group she is visiting.

If we consider the Quechua of the Peruvian Andes, we find a different way in which this relational self is understood—which can be traced to their cosmovision that represents all aspects of life as interdependent and in constant communication—with plants, animals, and weather patterns communicating what the people's agricultural decisions should be. The key point is that Lakoff and Johnson repeat the silences about the influence of culture that characterizes western philosophy. One of the consequences of perpetuating this early tradition of hubris is that readers who take them seriously are not likely to recognize that we have much to learn from cultures that have developed in ways that enabled them to live within the limits and possibilities of their bioregions.

While Lakoff and Johnson also remain silent about the ecological crises, the consensus of the world's scientists is that we are within a few generations of a tipping point when changes in human behaviors will no longer be able to slow the rate of global warming. It is important, therefore, to consider whether the Lakoff and Johnson theory of the embodied basis of metaphorical thinking is useful for understanding why the dominant western culture continues to promote an industrial/consumer-dependent lifestyle when the evidence continues to mount that it is ecologically unsustainable. The other question that needs to be raised is whether their theory of the embodied origins of our concepts can lead to fundamental changes in ways of thinking and behaviors that have a smaller ecological footprint—including changes in our policies of economic and cultural colonization that prevent other cultures from revitalizing their traditions of self-sufficiency and mutual support that are less dependent upon a money economy and that have a smaller ecological footprint.

The Lakoff and Johnson theory of metaphor fails to take into account that words have a culturally specific history and that the meaning of key words used today as the basis for understanding current problems and relationships were framed by the analogs established by earlier thinkers who were addressing issues in a different historical and cultural era. The analogs that framed the meaning of words such as "intelligence," "technology," "tradition," "individualism," "property," "freedom," "woman," "environment," and so forth, can be traced to earlier theories, powerful evocative experiences, and even to mythopoetic narratives such as found in the *Book of Genesis*. For example, the analogs that framed how the word "woman" was understood in the West over thousands of years did not arise out of the embodied/sensorimotor/neurally connected experience of individuals. Nor do today's widely accepted understandings, especially within the academic community, of such words and phrases as "tradition," "artificial intelligence," "property," and "enlightenment" have their origins in the subjective embodied experience of the individual. The analogs that continue to frame what is understood today as the meaning of these words can be traced to earlier events and influential thinkers. To summarize a key shortcoming of the Lakoff and Johnson theory about the origins of metaphorical thinking: it cannot account for the linguistic colonization of the present by the past. How the history of words is part of an ecology of language that takes account of the influence of other linguistic traditions is also beyond what they can explain.

Nor can their theory clarify the dynamics of the linguistic colonization of other cultures. Indeed, if one gives careful consideration to their six key assumptions, it becomes impossible to explain the differences in cultural ways of knowing—including why some cultures have developed in ways that are more ecologically sustainable. For example, the collection of essays by Third World writers in Wolfgang Sachs' *The Development Dictionary: A Guide to Knowledge as Power* (1992) provide examples of how such words as "development," "market," and "poverty"—whose meanings were framed by the analogs taken for granted by western thinkers—are understood as the language of cultural colonization. As Gerald Berthoud summarizes the colonizing agenda in the western use of "development":

> What must be universalized through development is a cultural complex centered around the notion that human life, if it is to be fully lived, cannot be constrained by limits of any kind. To produce such a result in traditional societies, for whom the supposedly primordial principle of boundless expansion in the technological and economic domains is generally alien, presupposes overcoming the symbolic and moral 'obstacles,' that is, ridding these societies of various inhibiting ideas and practices such as myths, ceremonies, rituals, networks of solidarity and the like.
>
> (Sachs, p. 72)

Berthoud identifies the analogs that frame the western idea of development by observing that it subjects Third World cultures to "the compelling idea that everything that can be made must be made, and then sold. Our universe (according to the western way of thinking) appears unshakably structured by the omnipotence of technoscientific truths and the laws of the market." (p. 71) The other essays in *The Development Dictionary* explain how key words in the modern vocabulary are not culturally neutral metaphors, but are part of the process of linguistic colonization that serves to legitimate economic colonization. And the combination of linguistic and economic colonization impacts the behavior of individuals—even at the level of the individual's "perceptual, imaging, and motor systems"—which is the reverse of the Lakoff and Johnson formula that represents bodily experiences as shaping the individual's conceptual system.

There is another characteristic of metaphorical thinking ignored by Lakoff and Johnson that is critical to whether we are able to adopt a more ecologically informed way of thinking and behaving. The image (or iconic) metaphors they focus on as originating in embodied experiences, as well as image metaphors that come down to us from earlier times (which they do not recognize) are often framed by the prevailing root metaphors of the culture. According to Richard H. Brown, root metaphors are meta-cognitive schemata that are taken for granted and thus frame thinking in a wide area of cultural activity over years—even centuries. (1977, p. 125) They originate in the mythopoetic narratives of the culture, in powerful evocative experiences that are sustained over generations, and from the processes of analog-based theories of writers who were able to overturn older root metaphors.

In the West, patriarchy and anthropocentrism are examples of taken for granted explanatory/moral frameworks (root metaphors) that have not only framed how people think and behave, but also what they ignore. Individualism and progress also are examples of root metaphors, and they can be traced back to various political theorists, and evocative experiences ranging from the introduction of the printing press to the early successes of modern science. Mechanism is yet another root metaphor whose origin was not in the individual's bodily experiences, but originated from a combination of historical events—ranging from organizing the rhythms of daily life in accordance with the cycles of a mechanical clock, the successful applications of a mechanistic paradigm by scientists, to advances in medicine and other technologies such as computers. The explanatory power of the mechanism root metaphor used over hundreds of years can be seen in Johannes Kepler's (1571–1630) statement that "my aim is to show that the celestial machine is to be likened not to a divine organism but to a clockwork;" in Marvin Minsky's (an early leader in the field of artificial intelligence) statement that "our conscious thoughts use signal-signs to steer the engines in our minds, controlling countless processes of which we're never much aware of;" in Richard Dawkins' reference to the body as a "survival machine;" in E. O. Wilson's reference to his brain as a machine; in the current way of identifying a plant cell as possessing a "powerhouse," "recycling center," and a "production center;" and in today's widespread references to the brain as being like a computer.

As meta-cognitive explanatory frameworks guide thought and behavior at a taken for granted level of consciousness, they exercise a profound influence on many aspects of culture—and thus on the embodied experiences of the individual. For example, the root metaphor of patriarchy established the analogs for understanding the identity and behavior of women in ways that were highly restrictive. It exercised this control over centuries until recently when the word "woman," in some sectors of society, became associated with a wide range of new analogs such as engineer, artist, doctor, politician, and so forth. We could take other mutually supportive root metaphors in the West and trace how they create areas of silence, limit the vocabulary to what is conceptually and morally coherent with the root metaphor, and thus control the discourse that frames how political problems are understood and the approach to resolving them. What is especially important about these western consciousness-shaping root metaphors is that they gave conceptual direction and moral legitimacy to the industrial/individually centered/consumer-dependent lifestyle that has been a major contributor to global warming and to the economic exploitation of the environment. As fisheries disappear, droughts become more widespread, storms more violent, and sources of potable water increasingly scarce and contaminated, the embodied experiences that Lakoff and Johnson want to claim as the source of concepts and inferences are unlikely to lead to an awareness that everyday life is being influenced by a symbolic ecology that is still being reinforced at every level of the educational process—and by the media, political pundits, and even by many environmentalists.

Embodied experience alone will not provide the conceptual and linguistic capital necessary for recognizing the double bind thinking that limits our ability to renew the intergenerational patterns of self-sufficiency and mutual support that represent alternatives to the industrial/consumer lifestyle that is moving us closer to the tipping point that will have huge consequences for the embodied experience of the individual—such as social chaos, starvation, and toxic chemical-caused illnesses and death. What is ironic, especially since the new root metaphor of evolution is now being used to explain how cultural patterns (memes) are subject to the same process of natural selection, is that it is being promoted by professors who are unaware that when evolution is turned into a root metaphor that supposedly explains the symbolic world of culture, it then supports the market liberal ideology that, in being globalized, is exacerbating the ecological crises.

While Lakoff and Johnson claim that "our conceptual system is grounded in, neurally makes use of, and is crucially shaped by our perceptual and motor systems," it turns out that their writings and ways of understanding political issues have been heavily influenced by the root metaphors they take for granted—and of which they are not explicitly aware. For example, Johnson responded in a letter to my earlier criticism of Lakoff's lack of historical accuracy and misuse of our political categories by claiming that there is nothing problematic with Lakoff's reference to environmentalists as progressives. The point to keep in focus is that Johnson's association of environmentalism with the forces of progress, and Lakoff's reliance on the same political language that underlies the cultural forces that are pushing the world beyond what the ecosystems can sustain, is that their concepts are not derived from their own embodied experiences. If they had done an ethnographic description of their own embodied experience (or what Clifford Geertz calls "thick description") they would have found that their use of the context-free vocabulary of freedom, individual autonomy, and linear progress are derived not from their own embodied experience, but from the reification of analogs derived from Enlightenment thinkers that are reproduced in the languaging/socialization process. In western cultures, the infant's first encounters with the ecologies of sounds, signs, and identities are quickly transformed and narrowed to fit the largely print-influenced abstract realities of adults, of the media, and of the formal educational process.

Lakoff, in particular, is unable to rely upon his own theory when it comes to justifying his political preferences and to stigmatizing his political opponents. In his *New York Times* best selling book, *Don't Think of an Elephant: Know Your Values and Frame the Debate* (2004), Lakoff makes an important contribution to understanding how the use of language frames what is given attention and what is marginalized in today's political discourse. He makes the point that language is not politically neutral. Instead, the group that is able to establish its preferred vocabulary (including its silences and metaphorically based prejudices) will control the policies that are conceptually consistent with its language. In effect, a group cannot achieve its own political goals if it is forced to think in the opponent's language.

Instead of following what is derived from his own embodied-based reasoning, Lakoff adopts the Orwellian vocabulary that is now current at every level of American political discourse. This discourse labels as conservatives the market liberals who derive their ideas about free markets and the invisible hand from the abstract theories of classical liberal thinkers—and the religious fundamentalists who derive their guiding principles from the equally abstract idea that the *Bible,* which has undergone many translations, represents the actual word of God. Lakoff's historically uninformed thinking leads to identifying as progressive the social groups concerned with conserving our civil liberties (the American Civil Liberties Union), the environmentalists working to conserve species and habitats, the people who translate their religious traditions into social justice activism, and ethnic groups working to sustain the connections between their identities and their traditions. (p. 14)

Lakoff also labels as conservatives the CATO, American Enterprise, and Hoover Institutes. If he had checked out their websites, rather than relying upon popular misconceptions, he would have found that all three identify the expansion of free markets, individual freedom, and a strong defense as their primary political agenda. He would have found the following on the website of the CATO Institute: "'Conservative' smacks of an unwillingness to change, of a desire to preserve the status quo. Only in America do people seem to refer to free-market capitalism—the most progressive, dynamic, and ever-changing system the world has ever known—as conservative." Lakoff is old enough to remember Ronald Reagan introducing the General Electric weekly television program in which the GE mantra of "Progress is our most important product" served as the analog the public was to identify with GE technologies. Surely, he is aware that the techno–scientific/industrial culture has always claimed the role of being the primary progressive force in society. It has only been in recent years that the ever-changing system of free-market capitalism has been labeled as conservative. (Hartz, 1955) The widespread ignorance that underlies this Orwellian misuse of our political language can be partly attributable to the failure of universities to introduce students to the history of current ideologies.

As pointed out earlier, one of the characteristics of a root metaphor is that its supporting vocabulary does not include the words that enable the basic taken for granted cultural assumptions upon which it rests to be questioned. In the case of the root metaphor of progress, the two words that are either proscribed or mis-represented are "tradition" and "conserving." By relying upon the root metaphor of progress to frame his analysis of liberalism and conservatism, Lakoff falls into the conceptual trap of letting the root metaphor of progress dictate what should not be questioned: namely, whether interpreting all forms of change as the expression of progress is partly responsible for undermining the traditions that should be conserved, such as habeas corpus, the Constitution, and the Bill of Rights. Since the early days of the 1900s, there has been a movement to conserve the environmental commons, which includes what remains of the natural systems that have not been taken over by private and corporate ownership and turned into market opportunities (which is considered by market liberals as progress).

The cultural commons, which both Lakoff and Johnson could have made the focus of their discussion of embodied knowledge, includes the intergenerational knowledge, skills, and mutually supportive relationships that enable people to live more community-centered and thus less money dependent and less environmentally destructive lives. Yet it is these communities that are sources of resistance to the market system that the CATO Institute celebrates as the engine of progress. In short, by uncritically accepting an interpretative framework (root metaphor) that can be traced back to the Enlightenment thinkers, Lakoff abandons his own prescriptions for how to account for "Embodied Reason." In lacking an historical knowledge of the origins of philosophical conservatism, which led to our checks and balance system of government, as well as of the abstract theories of classical liberalism, Lakoff unknowingly aligns himself with the environmentally destructive and cultural colonizing forces of the market liberals, who promote progress as though it is a law of nature.

What should be noted is that the Lakoff and Johnson theory of the embodied reason and the embodied origins of metaphorical thinking have implications that go beyond their inability to abide by their own guiding assumptions. The deeper problem is their lack of awareness that we are not only at a tipping point in terms of the rate of environmental changes, but also at a tipping point as to whether humankind can move beyond the myths that underlie the individualistic/consumer-oriented/industrial culture. The tipping point, in effect, involves the choice of following the current cultural agenda of economic globalization that is being adopted in many regions of the world, or revitalizing the local cultural and environmental commons that represent a post-industrial consciousness based on the ancient Greek root metaphor of *oikos*. Fortunately, many environmentally oriented scientists have moved beyond the mechanistic root metaphor by learning to think of the natural world as living ecologies, and many Third World cultures have not entirely lost their ecologically informed traditions.

The main challenge will be for philosophers and social theorists such as Lakoff and Johnson to explain the dangers of accepting without question the root metaphors that were constituted before there was an awareness of ecological limits and to explain, in ways that can be widely understood, how our everyday vocabulary in the West needs to be framed by analogs that are culturally and ecologically informed. Given their unquestioning embrace of the root metaphor of progress and their commitment to assuming that embodied experience is the primary source of the metaphors that guide thought and behaviors, it is doubtful that their contribution will be little more than yet another distraction as we move closer to the ecological tipping point. Perhaps if they were to start not with the embodied experience of the supposedly autonomous individual, but with the individual's culturally mediated embodied experience, they would have the conceptual opening for considering the influence of culture, the role of language in the cultural construction of identities and ways of thinking, the influence of diverse cultures on embodied experiences, and the ecological implications of doing a "thick description" or personal ethnography of the interdependencies within the local cultural commons. This would have represented a genuine departure from the silences and hubris of most western philosophers.

References

Bateson, Gregory 1972. *Steps to an Ecology of Mind.* New York: Ballantine Books.

Brown, Lester. 2008. *Plan B 3.0: Mobilizing to Save Civilization.* New York: W.W. Norton.

Brown, Richard H. 1977. *A Poetic for Sociology: Toward a Logic of Discovery for the Human Sciences.* Cambridge: Cambridge University Press.

Dawkins, Richard. 1976. *The Selfish Gene.* New York: Oxford University Press.

Geertz, Clifford. 1973. *Interpretations of Cultures.* New York: Basic Books.

Hartz, Louis. 1955. *The Liberal Tradition in America: An Interpretation of American Political Thought Since the Revolution.* New York: Harcourt, Brace.

Johnson, Mark. 2007. *The Meaning of the Body: Aesthetics of Human Understanding.* Chicago: University of Chicago Press.

Lakoff, George. 2004. *Don't Think of an Elephant: Know Your Values and Frame the Debate.* White River Junction, VT: Chelsea Green.

Lakoff, George, and Mark Johnson. 1980. *Metaphors We Live By.* Chicago: University of Chicago Press.

Lakoff, George, and Mark Johnson. 1999. *Philosophy in the Flesh: The Embodied Mind and Its Challenge to Western Thought.* New York: Basic Books.

Sachs, Wolfgang. 1992. *The Development Dictionary: A Guide to Knowledge as Power.* London. Zed Books.

6 The Cultural Construction and Uses of Data

Everyday life relies upon misconceptions carried forward from the past, while encountering new challenges that are only superficially understood. The current emphasis on data-based decision-making represents one of these double-bind situations, where the more we rely upon past misconceptions about the nature of data the more we will increase our superficial understanding of the challenges we face.

The misconceptions inherited from the past had their origins in what were then progressive insights of how to free thinking from the superstitions of the day. These progressive insights included relying upon critical inquiry, the rationalism of the individual, and the new mode of inquiry where empirical evidence was needed in order to establish a new form of knowledge: namely, objective knowledge. The ability to use this knowledge to bring about changes in the world required that it met the test of being replicated by others who also elevated the importance of empirical evidence in determining what constituted objective knowledge that was free of both the prejudices of the investigators and the prevailing myths of the larger society.

This new mode of inquiry became known as the scientific method, and it represented a real break from both the taken-for-granted world of widely held superstitions as well as the abstract theories of mainstream philosophers and social theorists of the times. Its success in giving control and predictability in bringing about changes that improved the quality of life led to a new set of taken-for-granted beliefs, including the idea that there is such a thing as objective knowledge acquired through a mode of inquiry that is free of cultural influences. Actually, in the late 16th and early 17th century when this new mode of inquiry was becoming established, there was no understanding of culture.

More importantly, there was no understanding of the role that language plays in reproducing the thinking of earlier eras—and how the vocabularies inherited from the past influence awareness, interpretations, as well as what is not recognized. In short, the origins of modern science did not take account of the ontological reality that humans cannot escape how the scientists and everyone else takes for granted the linguistic influences on their interpretations of the world. That is, the scientific gaze, while focused on evidence requiring that hypotheses meet the test of verifiability by other scientists, cannot entirely escape the influence of how the taken-for-granted language frames how the world is interpreted.

That the world of scientists is an interpreted world, one that takes account of empirical and measurable evidence, can be seen in how their discoveries are viewed within western cultures as expressions of a linear form of progress. The scientific discoveries that promoted the development of the first indus- trial and now digital revolution were justified as leading to a more progressive future—even though we now recognize that the real progress was in contrib- uting to what we now understand as climate change and the acidification of the world's oceans. Scientists in Nazi Germany took for granted the social Darwinism that justified the elimination of the less fit humans, just as scien- tists supported the eugenics movements in North America and Great Britain. Their interpreted world also supported the idea of equating intelligence with learning to think in the English language. And now, reducing the complexity of everyday life experiences to data, the ability to collect and store vast quanti- ties of data on people's behaviors and ideas, and developing the technologies that now subject everyone to being hacked, are interpreted as further expres- sions of progress. Indeed, it is the scientific mindset that reduces the semiotic worlds of communication that sustain both the natural and cultural emergent, relational, and co-dependent ecologies to what can be observed through the use of MRI technologies. Increasingly, measurability that can be represented as data is becoming a key criterion for determining if something exists.

The historical developments in how the scientific gaze is now reducing the inter-subjective world of personal identities and narratives to the objective status of data now has to be reconciled with how we are beginning to under- stand the ontological realities of the world we live in. These realities include the following: 1) that there is no permanence in either the natural or cultural ecologies. Everything is emergent according to its cycle of renewal, which is influenced by a second ontological characteristic. 2) That is, everything exists in relationship with other participants in the natural and cultural ecolo- gies. Contrary to a major misconception in the west, there are no isolated, self-contained entities. To put it another way, there is no autonomy such as autonomous thinkers and actors. The relational nature of existence needs to be understood as ecologies of communication where everything from the simplest organism to the most complex natural and cultural systems relies upon its inherited semiotic system for interacting and influencing the Other. 3) The ontology of the world we live in also includes the co-dependence that is sustained through the ongoing semiotic patterns of communication (or information exchanges) that represent how each organism and natural pro- cess respond to the emergent nature of the ecological world that they are part of.

In the case of the west, other misconceptions reproduced in the inherited vocabularies include the idea that this is a human-centered world, that there is such a thing as an autonomous individual, and that a consumer-based lifestyle leads to progress.

Influence of Language on Awareness and Interpretations

As the digital culture now transforms more aspects of the culture of everyday life into data, the old misconceptions continue to be carried forward—even by scientists. And the primary misconception is that data is objective and measurable—to be understood and used as factual information free of cultural influences. The myth of progress is also a cultural construction that has been justified in terms of different narratives, such as the Book of Genesis, the abstract thinking of western philosophers such as John Locke and René Descartes who argued that reliance upon their different epistemologies would free people from the constraints of traditions, and now the computer futurists who are racing to replace human intelligence with artificial intelligence— even as they are blind to the reality that intelligence is largely cultural and thus differs between cultures.

Data and print share the same basic limitations. And like print, which in the west has been understood as mostly responsible for its many cultural achievements, data is now being seen as leading to further progress. The following limitations of print are also the limitations of data:

a   Data, like print, can only provide a surface knowledge of ideas, events, and processes. It cannot fully reproduce contexts.
b   Data, like print, immediately becomes outdated in the ecological world of constantly changing relationships and multiple levels of message exchanges.
c   The abstract thinking fostered by data and print too often becomes interpreted as having a universal meaning.
d   Both data and print reinforce the conduit view of language that undermines awareness that words are metaphors that carry forward earlier cultural assumptions that influence current interpretations.
e   Both data and print reinforce the western myth of the autonomous individual who relies primarily upon a visual relationship with the external world.
f   Both data and print are inherently ethnocentric as they are unable to represent the inter-subjective world of oral cultures.
g   Most writers and readers are unaware of the taken-for-granted cultural assumptions that influence their interpretations of the world that take on the appearance of objectivity when encoded in the printed word and as data.

But data is not just acquired through measurement and constant surveillance of behaviors in the natural and cultural ecologies. Decisions about what in the diversity of ecologies that is to be taken out of context, reduced to a surface snapshot of phenomena in the emergent world of natural and cultural ecologies, is driven by powerful interpretative frameworks generally taken for granted by the person or group that decides on what needs to be transformed into data. The interpretative framework or ideology of the scientists collecting data on changes in water temperature or rate of acidification of a local body of water includes both the scientific method but also assumptions about the need to slow the rate of environmental degradation. And the culture's emphasis that change be interpreted within its calculus of progress is also an inescapable part of the scientist's thinking. The educator collecting data on a student's performance, as well as the manager of a hospital, factory, bank, and so forth who wants data on the performance of workers, relies upon an interpretive (ideology) framework that emphasizes a different set of cultural priorities such as the importance of improved productivity, greater efficiency, improved profits, and so forth.

The interpretive framework, in effect, dictates what is to be collected, and provides the justification for ignoring the complexity of an ecology of human experience and of other phenomena in the natural world. Like print, data requires that the emergent, relational, and co-dependence that are characteristic of all natural and cultural contexts be ignored. Also, like print, where the writer is often unaware that the vocabulary she/he takes for granted encodes the metaphorical thinking of earlier eras, the human/cultural influences on the decision-making process about what constitutes data are lost sight of. Reduced to numbers, words, graphs, and computer models, the human authorship disappears. And when this process of reductionism is understood as an expression of the scientific gaze, few people are likely to question the cultural roots of this process—or the human costs. It is much easier to assume that the data possesses an objective status.

Again, it is important to recognize that, like print, there are many benefits in acquiring data on the behavior of different systems—especially in light of climate change and the many ways the world's natural systems are now being stressed. But the collection of data can also lead to new forms of social injustice, such as what is occurring in the digital revolution where the dominant interpretive framework (ideology) is driven by the idea that progress requires that computer intelligence (AI systems) replace human intelligence, as well as human workers—and where collecting data on people's behaviors is more important than protecting their privacy.

The introduction of the Internet, and now the increasingly digitally connected world of everything, which will lead to even more massive amounts of data that will be stored in the cloud, by-passed the democratic process in the first place. And now in, the name of technological progress, the Internet is introducing new perils into the world where cyber warfare is becoming an increasing threat as nations continue to compete for resources, markets, and influence on the world stage. There is also an increased threat that what remains of the democratic process will be further eroded as national security concerns, as well as criminal and terrorist networks, rely more upon surveillance technologies that, in turn, lead to increasing the police powers within society.

That the digital revolution, for all of its benefits—and there are many—is leading us down the pathway to a techno-fascist future raises an important problem that is just beginning to be recognized, which in itself raises the question of why the current concern of computer experts with identifying a guiding ethic has just come up. It would seem that a concern with the morally inappropriate uses of digital technologies would have been the initial concern before the technologies that now make everyone, everywhere in the world, vulnerable to being hacked, before algorithms that now replace workers were introduced, and before the total surveillance systems were put in place. The social justice issues should have been addressed at the outset, as it is now impossible to limit hackers, extremists engaged in cyber attacks, and governments and businesses that benefit from the surveillance culture that is emerging.

What is seldom recognized is that when an individual's performance, relationships, and ideas have been reduced to data, and can be accessed by others, the individual has no control over how the data will be used. The ideology of the Other then becomes part of the ecology of data collection and use. The Other may be a well-intentioned person driven by a personal sense of integrity, but she/he could also be driven by the profit motive, by a desire to make the lives of others more difficult, to bullying, and generally to exploiting the vulnerabilities of others. All the safeguards that have been created in the west to protect the civil rights of the individual can now be overcome, with the state no longer providing protections. The digital revolution has in effect marginalized both the authority and ability of the state to protect its citizens.

Computer Mediated Learning and the Loss of Knowledge of Local Contexts and the Lived Experience of Others

The word "context" is another abstraction, but it refers to the ecology of experience where the emotional, intuitive, empathic, commemorative, reflective, spiritual, and other ongoing negotiations with the Others reproduce taken-for-granted cultural patterns in this emergent, relational, and co-dependent world. Each individual's cultural/existential context is biographically distinct, even as many of the cultural patterns frame how the existential aspects of personal experience are expressed. These existential, that is, personal, sources of meanings have a history, are largely framed by cultural/linguistic influences that are taken for granted, and may be the sources of self-doubt, sense of inferiority, and confusion about life's purpose.

These ecologies of how relationships are experienced in this emergent world are impacted by the decisions of others who have been educated to think in abstractions, to create social systems that are intended to fit people's lives in pre-determined patterns—often in ways that differ radically from the cultural patterns that have become a taken-for-granted lifestyle. The Others who are now creating the learning environments where youth are learning to think in the abstractions created by the printed words appearing on the computer screen, and in terms of data, graphs, and computer models of how some part of the world works, are being socialized to ignore both their own inter-subjective worlds as well as the inter-subjective worlds of others—including the culturally different. What they are learning to take for granted is that their own ecologies of experience can be accurately represented as data that others will use, and that carry forward the existential moments in time when the data were collected. The data, as pointed out before, represents fixed moments in the flow of time, while the life-worlds of living experience are always emerging even as the patterns of thinking and behavior continue to be influenced by cultural traditions that are taken for granted.

One of the characteristics of both the print appearing on the computer screen, and the data that reflects the abstract and thus surface thinking of the people who created the curriculum, provides a model of thinking that can heavily influence the thinking of a young student who is learning something for the first time. For students who are already being unconsciously influenced by the mindset of their teachers and curriculum developers, encountering the same mindset over and over again further ensures that it will be experienced as the natural order of reality even when it only represents a surface, abstract, and ideologically driven interpretation of reality.

There are a number of characteristics of this print and digitally constructed process of learning that affects consciousness itself. One consequence is that attention spans are shortened, which makes it less likely that thinking will encounter the depth of explanations—which includes the historical and the mix of events and information that influenced the development of earlier or current times. Another consequence of a print and data-dominated curriculum is that in-depth learning about other cultures, particularly oral cultures, will be limited—and too often represented as backward because they value the ecologies of face-to-face relationships rather than the abstractions of the print/data-based world. This will lead to cultures that have developed ecological intelligence, which comes from giving close attention to the emergent, relational, and co-dependent patterns occurring in natural systems, being ignored.

Another consequence is that students will not learn about the many ways that language is partly at the root of the ecological crisis. That is, they are unlikely to learn how vocabularies encode and thus carry forward many of the misconceptions of earlier eras when the meanings of words were being framed by the choice of analogs that reflected the misconceptions and silences of the times. John Locke settled upon the analog of human labor as determining the basis of property ownership, while Adam Smith, in promoting free markets, helped to undermine the long tradition of the guild systems. We can clearly see now the biases and misconceptions that framed the early meaning of the words "woman," "environment," "individualism," "intelligence," and that this is a human-centered world. In being unaware that words have a culturally distinct history, words then are easily misunderstood as referring to things and events that are real in themselves and not cultural constructions that continue to frame the current thinking that ignores climate change and other evidence showing that we may be entering the extinction of the world species—including ourselves.

For the computer scientists who are influenced by science and the myth of objective knowledge, rather than by a deep knowledge of cultural differences, the current efforts to globalize the digital revolution will continue. This process of cultural imperialism will also continue to replace the intergenerational knowledge and skills essential to the formation of personal identities and loyalties with a data-based view of the world. This, in turn, opens the door to replacing culturally diverse humans with the new digital technologies that will increase the efficiencies in the economy that is now threatening the world with ecological collapse.

Part II
The Cultural Commons

7 The Political Economy of the Cultural Commons and the Nature of Sustainable Wealth

People of all ages are awakening to a reality that has been hidden by years of seemingly limitless consumerism and the expectation of lifetime employment. This has been an evolving reality marked by increased automation, caution-to-the-wind expansion of manufacturing capacity, outsourcing of jobs to low-wage regions of the world, the breakdown in the social contract between employers and employees, and the increasing sense of entitlement that gives the heads of corporations the right to millions of dollars in compensation regardless of their performance. The consequences of these largely ignored realities are now affecting the lives of both students and adults. Unemployment and working for a minimum wage (if that is even available), the threat of losing one's home to foreclosure, the inability to pay for health care, growing food insecurity, and the reduced hopes for further education are the realities now experienced by millions of people. To rework Charles Dickens' famous phrase, the best of times are now turning into the worst of times.

The spread of poverty continues with little hope in sight, especially now that fear is replacing the myth of unending progress in accumulating more material wealth. The fear, and the sense of helplessness that accompanies it, are based on years of being socialized by the media and other consciousness-shaping forces to equate wealth with gains in the money economy. In short, the amount of money one acquires has become the primary measure of wealth. This narrow understanding of wealth has led to a competitive form of politics in which the achievement of success requires placing one's own economic interests over the well-being of others. Individuals, families, and ethnic groups gained in wealth as they took advantage of the marginalized and thus politically powerless, just as the wealth of corporations depended upon paying workers as little as possible. Indeed, the more economically vulnerable the workers, the more easily they can be underpaid and their past gains in the workplace revoked. The role of government, as many market liberals understand it today, is not to impose limitations on industrial capitalism, while being ever-ready to pass legislation that furthers the interests of corporate lobbyists who provide the money necessary for winning elections. In short, the form of politics essential to an industrial/market/consumer economy operates behind the façade of being democratic, but it continues to be based on the competitive pursuit of self-interest in which money determines, with few exceptions, who will be the winners and losers in achieving even greater material wealth.

This form of politics and the pursuit of profits will also ensure that the fate of natural systems will continue to be an "exploitable resource." It must be acknowledged, however, that there is an increased awareness that environments are being degraded in ways that will further diminish the material wealth of this and future generations. This awareness is now creating greater tensions between political factions. Unfortunately, nearly half of the voting public still thinks of the free-market ideology, and its underlying assumptions, as having the same status as the law of gravity. For the majority of these followers of Adam Smith (who badly misinterpreted his ideas), Milton Friedman, and today's libertarians, environmental changes are part of the natural cycles that have occurred over millions of years and cannot be attributed to the excesses of human behavior.

The failure of public schools and universities to challenge the dominant cultural assumptions that underlie the political and economic system that equates wealth with the possession of money, and the credential system that provides access to power and money, have left most people ignorant of how to avoid sinking further into poverty and hopelessness. Part of the failure of these institutions, which is reproduced by their graduates who use the media to promote the same misconceptions and silences acquired as part of their university education, is in not introducing students to a more complex and community-grounded understanding of the sustainable forms of wealth that represent alternatives to what is dependent upon the money economy. This failure is partly linguistic, partly rooted in the high-status accorded to abstract knowledge and patterns of thinking, and partly rooted in a combination of cultural developments connected with the rise of science and what has become the mythic foundations of modernity. These mythic foundations include the idea of the autonomous individual, the progressive nature of change, the culture-free nature of the national process and critical inquiry, an anthropocentric view of human/nature relationships, and the Darwinian view that the competitive nature of free markets will determine which genes and cultural memes are best fit to survive. It is important, however, to recognize that not all members of local communities or ethnic groups in North America have based their lives on these assumptions. Indeed, many have discovered the non-monetized forms of wealth that have largely been ignored in the curricula of public schools and universities.

These non-monetized forms of wealth have not only been important to sustaining the lives of people locked out of the money economy, but they are also taken for granted by people who live well above the poverty line. Because these non-monetized forms of wealth have been accorded low status and thus omitted from the curricula of most public schools and universities, graduates are caught in a double-bind of which they are not aware. They lack an explicit awareness of the non-monetized community and inter-generational forms of wealth they rely upon in most daily activities, while at the same time they look forward to a return to the days of unbridled consumerism and life-time employment. The reality they will encounter in the future will be quite different. Automation, outsourcing, and downsizing are here to stay. In addition, the primary need of the industrial system of production and consumption to expand will lead to turning more of the non-monetized relationships and activities into new market opportunities, thus further increasing people's dependence upon the money economy. Because of the historical roots of this system of production, and the cultural assumptions upon which it is based, it has not occurred to most people that the individualistic, competitive, consumer-dependent lifestyle, and its accompanying form of politics, are not ecologically sustainable—even over the short run. This double-bind is more than a short-lived down turn in the economy. It now characterizes the embodied experiences of millions of people who seek a return to the halcyon days but are unable to recognize that those days will not return. Even more important is that they are unable to recognize the alternative forms of wealth that are part of the cultural commons of every community.

The task here is to clarify the forms of wealth intrinsic to the cultural commons, including how they differ from the wealth acquired by participating in the money economy. Money is useful in many ways, and it will continue to have a role to play in facilitating exchanges in the larger world. Its role, however, will be reduced by environmental as well as by global technological and cultural changes. These changes may range from Third World cultures resisting the western model of development to the collapse of the modern state that we are now witnessing in some regions of the world. Thus, it is now imperative that people obtain an explicit understanding of the unique characteristics of the wealth that is available in the local cultural commons. The wealth of the cultural commons takes many forms and has the following unique characteristics. It enables people to discover interests and talents that lead to less stressful and thus less debilitating lives, to lifestyles that have a smaller adverse impact on the ability of natural systems to renew themselves, to alternative ways of reducing dependence upon processed foods that are costly and often unhealthy, and to maintaining the local traditions of participatory decision making that safeguard against the further integration into the market economy of what remains of the local cultural and environmental commons.

A second task is to clarify the form of politics that supports the alternative economies of the cultural commons that vary from culture to culture. This task may seem rather straightforward, but it needs to be recognized that hundreds of years of mis-education are responsible for the difficulty many people have in being explicitly aware of the nature of their local cultural commons—even as they tacitly rely upon them as part of their everyday lives. There are also the problems of misinterpretation in which readers will reach conclusions that reflect their own unexamined taken for granted assumptions and, in some cases, romanticize the idea of the cultural commons rather than recognize examples of the cultural commons that do not fit current norms of social and ecological justice. There is also the challenge of introducing new ways of understanding the meaning of words, as well as recognizing that words such as "wealth" and "commons" have different meanings in different cultures and in different historical periods in the West. Hopefully, these sources of resistance will not hamper efforts to consider the educational reform implications of introducing students to the political economy of the cultural commons, or the policy issues required to achieve a better balance between participating in the money economy and the lifestyles that are more engaged in renewing the cultural, and by extension, the environmental commons.

In order to understand how the cultural commons represent alternative forms of wealth, it is necessary to go beyond abstract descriptions. Abstract descriptions found in printed texts too often are reduced to identifying what turns out to be general categories of intergenerational knowledge, skills, and mutually supportive relationships—such as the growing, preparing, and sharing of food, knowledge of the medicinal characteristics of plants and traditionally proven remedies, narratives and ceremonies, forms of artistic expression and craft knowledge, rules and practices that govern moral relationships and forms of reciprocity, knowledge of how to adapt cultural practices to the life cycles that sustain local ecosystems, and so forth. Each of these categories needs to be understood in terms of culturally diverse local practices and, more importantly, the depth of background knowledge that the activities in each of these categories depend upon.

In order to grasp a partial understanding of how different aspects of the local cultural commons are dependent upon the accumulated intergenerational knowledge and skills, it is necessary to do an auto-ethnographic account of how different aspects of the cultural commons are the basis of daily experience. Examples might include a description of how using recipes passed down within the family or through widely shared cultural practices are dependent upon knowledge gained and refined in the past. The auto-ethnography might focus on the background knowledge and intergenerational traditions that now lead to the taken for granted expectation that one's home will not be searched by government agents without a search warrant. Reliance upon proven techniques in framing the walls of a house, playing a piano, and following the rules of a game are also examples of intergenerational wealth that is the source of individual and group empowerment.

Admittedly, it is difficult to do an auto-ethnography of the layers of accumulated knowledge and skills that are relied upon when participating in the cultural commons that we casually refer to as everyday life experiences. We are too often absorbed in completing the task at hand, and moving on to another task, to consider the knowledge and skills accumulated over many generations that we tacitly rely upon. The fast pace required by the increasing dependence upon technology and the need to participate in the cycle of work, consuming, and meeting debt payments, contributes to a permanent state of cultural amnesia. Perhaps the even more overriding reason for the current state of ignorance of the wealth of the cultural commons is that it is largely taken for granted. Thus, what is taken for granted is often the tacit knowledge of skills, values, and activities that are relied upon in different physical and cultural contexts.

Unfortunately, when explicit awareness of the different forms of intergenerational knowledge and skills is lacking, outside economic and political forces may undermine or appropriate different aspects of the cultural commons without people knowing what has been lost. For example, important parts of our vocabulary have been lost to the forces of science and technology, just as non-western cultures have lost traditions of intergenerational knowledge as their youth have been socialized to adopt the western assumptions essential to making them dependent upon an industrial/consumer-dependent lifestyle. Socializing the poor to the values and vocabulary that support dependence upon processed food, as well as the loss of intergenerational knowledge, has further undermined their health when they could more easily have afforded the basic ingredients that previous generations relied upon for a healthy diet. Examples of how not being aware of the wealth of intergenerational knowledge that represents alternatives to dependence upon the industrial/consumer-dependent lifestyle contributes to poverty and helplessness can be multiplied many times over.

Key Characteristics of the Wealth of the Cultural Commons

A primary characteristic among the diversity of the world's cultural commons is that the wealth of the cultural commons cannot be put in the bank, invested in the stock market, or limited to a privileged few. Rather, it exists as the source of empowerment in daily practices, ways of thinking, patterns of moral reciprocity, as a source of self-confidence, as knowledge of what practices and policies have proven dangerous to life and community, as the accumulated knowledge and technical skill that lies behind every major advance in knowledge, social justice, and technology. Potentially, it is the most democratic form of wealth, as it is shared through conversations, mentoring, and observing others, as well as through narratives and the arts. Learning to think and communicate in the languaging processes of the community is the first step in acquiring the accumulated wealth of the cultural commons. As participating in the cultural commons involves actions, performances, and relationships, it makes more sense to think of the language describing the cultural commons as verbs rather than as nouns that represent it as an abstraction and an object of analysis. Another characteristic of the accumulated knowledge, skills, and moral wisdom that is integral to many cultural commons is that as a form of wealth it cannot be lost through inflation or affected by the cycles of a money economy.

Indeed, as reliance on the money economy is threatened by the various excesses of greed, consumer debt, over production, and collapsing markets, people become more aware of the need to rely on the wealth of cultural commons. The recent collapse of the economic system in Iceland is a prime example. As the source of money and employment dried up as a result of the failures occurring in the national and international banking systems, the people turned to sources of wealth that were part of their cultural heritage. That is, they turned to the wealth of their cultural commons. Instead of importing goods and services, the people of Reykjavik turned to the knowledge and skills passed down by their grandparents, who were themselves inheritors of the accumulated wealth of earlier generations. Instead of the descent into poverty, the people began to rely upon the wealth of knowledge that enabled them to create from wood, metal, and fabrics items that could be exchanged and sold locally.

The current breakdown in the market economy has led to a similar recognition of the importance of the knowledge and skills of the local cultural commons. This includes the increase in the number of individual and community gardens, the revival of interest in various crafts, and the increase in volunteerism that in some communities has risen to over 36 percent of the local population and is focused on human needs ranging from food, repairing dwellings, and restoration of local ecosystems to community performances. Local markets, as well as a revival of bartering and the use of local currencies, are also part of the turn toward greater reliance upon the wealth of the local cultural and environmental commons.

This revitalization of the cultural commons is only a minor trend occurring across the nation and does not yet represent a major shift in consciousness. The majority of Americans, even in being unemployed and facing foreclosure, are still hoping for a return to the days of a consumer-dependent lifestyle and to taking their chances on achieving success in a money-dominated economy. This expectation is being reinforced by politicians who are continuing to promise a return to the lifestyle required by the industrial system of production and consumption, even as they also warn that the deepening ecological crisis will require new advances in technology.

To obtain a fuller understanding of how people are dependent upon the wealth of the local cultural commons, even during years of a growing economy, it is necessary to consider what represents inherited knowledge and skill and what is original to the individual. Does the craftsperson who is making a cabinet, or a violin, or framing the opening for a window, rely entirely on what she/he originates? Is knowledge acquired through trial and error of how to make the corners of a drawer that are both aesthetically pleasing and strong, or is it more often learned through a mentoring relationship, by following the advice of a neighbor or family member—or even following a manual? Did the Jonas Salks and Albert Einsteins of the world rely upon the accumulated wealth of the cultural commons of which they were members in order to make their discoveries? In short, are there examples of cutting-edge technologies or systems of thinking that do not depend upon a shared heritage? On a less lofty level, the craftsperson making a musical instrument is empowered when she/he can draw upon the knowledge accumulated by earlier generations about the sounds that will resonate from the use of different woods. Similarly, learning the rules of a chess game, the soil conditions and length of growing seasons for different plants, the way to prepare a curry and to preserve food, the patterns of meta-communication, and the established procedures to follow when one's civil rights have been violated are everyday examples of the widespread reliance on the shared wealth of the cultural commons. Sharing is essential to intergenerational renewal and is another characteristic that separates the wealth of the cultural commons from what is privately owned.

While vast amounts of information (much of it abstract and thus taken out of context) is increasingly available on the Internet, it is nevertheless different from the knowledge and skills passed on through face-to-face communication. When the wealth of the commons is encoded digitally it does not take account of cultural contexts, tacit understandings, and the powerful learning experience shaped by patterns of meta-communication that are part of mentoring relationships. Turning the wealth of the cultural commons into abstract descriptions has certain advantages, but it is also the first step to turning it into a monetized commodity. It is also an important step toward the enclosure of the cultural commons.

Before turning to a closer examination of the various forms of enclosure that students need to understand if they are to participate politically as adults in making decisions about what aspects of the local cultural commons need to be intergenerationally renewed, and which need to be modified or abandoned entirely, it is necessary to recognize that many cultural commons carry on traditions that are sources of exploitation and oppression. That is, the heritage or what is being referred to here as intergenerational knowledge may be a mix of wisdom of how to meet certain basic needs as well as prejudices that perpetuate various forms of discrimination and unjust social practices. For example, there are regions in the United States that have highly developed community-centered musical traditions (an important form of wealth), while at the same time they carry on traditions of racial and gender discrimination. These forms of discrimination lead, in turn, to reduced opportunities to participate in the money economy at a level necessary for meeting basic food, shelter, medical, and educational needs.

Summary of the Differences between the Political Economy of the Cultural Commons and the Market/ Industrial System of Production and Consumption

Focusing on the politics that separate the two economies brings out fundamental differences. A key difference is that the politics of many cultural commons are democratic in a way that empowers the community's traditions of mutual support and self-sufficiency. As skills and knowledge are shared in face-to-face relationships, and through other forms of intergenerational communication, questions and insights are shared. In effect, the interpersonal politics involve the element of mutuality and respect for others, which is at the core of Martin Buber's description of dialogue. It is the form of the politics found in mentoring relationships—though, to be realistic, mentoring is not always free of petty and even intergenerational misunderstandings. The politics of the cultural commons can also be seen in the distinction that Guillermo Bonfil Batalla makes between a culture where the norm is returning work as opposed to paying for work. (1996) The former is the politics of mutual support, while the latter is too often the politics of self-interest. There may be social hierarchies and systems of exclusions that influence who shares in the wealth of the cultural and environmental commons. These are sources of injustice and social pathologies that need to be overcome. In the healthy and life-enhancing aspects of the local cultural commons, wealth is found in sustaining the diversity of talents and skills, and in maintaining the intergenerational connections.

The politics of the industrial/market economy are profoundly different. In these economies, there is an emphasis on private ownership, and on accumulating more wealth—which is often achieved by reducing the opportunities and wages of others. In addition, the dominant ethos is to reduce the role of workers in making decisions about the process of work and the overall goals of the business. Competition rather than cooperation governs most relationships. Exploiting the human vulnerabilities of wanting to consume what is stylish and conveys a higher status in the community also figures into the politics of the industrial/market economy. At the governmental level, lobbyists pour vast sums of money into acquiring special advantages—which often take the form of obtaining tax breaks and huge government subsidies.

There is an even more destructive side to the politics practiced within the industrial/market sub-culture. This is the politics of enclosing as many aspects of the cultural commons as possible. This can be seen in how intergenerational approaches to meeting everyday needs ranging from food, healing, creative arts, craft knowledge, ceremonies, civil liberties, and so forth, are being turned into commodities and services that require participating in the money economy. The politics of enclosure may occur behind the façade of democratic decision making when the members of the local community have been indoctrinated to equate social progress with expanding the money economy and market. An educational system that represents the face-to-face, non-monetized intergenerational knowledge and skills as low-status and leaves them out of the curriculum, while representing the forms of knowledge required by the industrial/market-oriented culture as high-status, undermines the possibility of genuine democracy. For example, the silences perpetuated by public schools and universities about the wealth of knowledge that is part of our tradition of civil liberties easily leads to the kind of politics leading to fascism. Both youth and adults will be more welcoming of the latest technologies when the silences and accompanying prejudices falsely represent traditions as obstructing progress. As people become addicted to relying upon these technologies for communicating with others on a non-face-to-face basis, their lives become more hurried and stressful. This, in turn, leads to greater dependence upon the pharmaceutical industry to substitute their drugs and definitions of illness for the wealth of knowledge accumulated as part of the cultural commons of many cultures. As Vandana Shiva points out, many of the drugs that lead to vast profits are pirated from the intergenerational knowledge of indigenous cultures. (1993)

The following qualifications need to be kept in mind before addressing the educational reforms that enable students to share in the non-monetized wealth of their local cultural commons. We are now witnessing the monetized wealth that individuals and corporations invested in retirement accounts, bonds, stocks, and bank accounts losing value and largely disappearing. The intergenerational forms of wealth of the cultural commons may also be lost, especially when the prevailing ways of thinking are focused on the latest innovations and forms of entertainment. Examples that come readily to mind include how reliance upon industrially prepared food leads to a loss of knowledge of how to use traditional recipes to prepare a meal and to grow vegetables. As youth spend more time playing video games and participating in electronically driven social networks, there is less likelihood they will know the stories of their ancestors' achievements and wrongs done to others. Listening to market liberal talk show hosts such as Rush Limbaugh and the Fox News commentators will further undermine awareness of the accumulated political wisdom encoded in the U.S. Constitution, the Bill of Rights, and the gains in civil rights and social justice. The continual effort to expand markets in the name of progress also contributes to the further attrition of the cultural commons. The current lack of moral limits on what aspects of the cultural commons can be transformed into a commodity or service means that they are all under constant threat.

We should not think of the wealth of the cultural commons as entirely replacing the need for meaningful employment and a wage that enables people to meet their basic needs for food, shelter, health care, and education. Money is still required to purchase the goods and services that represent the genuine achievements of the scientific/industrial culture. However, the need for community, self-expression, and growth in developing an ecological form of intelligence can be met more fully by involvement in the local cultural commons. It's not an either/or issue, but one of balance that takes account of the excesses and exploitive nature of the industrial/consumer-oriented culture, as well as the need to live in ways that have a smaller adverse impact on the Earth's ecosystems.

What Students should Learn about the Differences between the Political Economy of the Cultural Commons and of the Free-Market System of Production and Consumption

The basic concepts that teachers and professors need to introduce include the following:

- The fundamental insight that should frame the discussion of educational reforms is Herman E. Daly's (1991) observation that while the environment establishes absolute limits on how far the industrial economy can expand, there are no environmental limits on the development of a culture's symbolic systems (or what is being referred to here as the life-and community-enhancing cultural commons).
- An auto-ethnography needs to be undertaken, as most aspects of the local cultural commons are experienced at a taken for granted level of awareness. This will involve a careful mapping of the intergenerational knowledge and skills that exist within the community, as well as the mentors who keep the traditions alive. This will ensure that the discussion is grounded in the culturally influenced embodied experiences of the students—and not treated as an abstract textbook explanation with which few students will be able to relate.
- A survey of the number of people who are living lives of voluntary simplicity, as well as those who are unemployed, under employed, and retired, needs to be undertaken, along with a survey of the knowledge that people have about the alternatives to meeting daily needs through consumerism.
- Initiate a discussion of how the wealth of the cultural commons differs from wealth in a money economy. This discussion should also include issues related to which forms of wealth are a human right and which have to be earned in settings where equality of opportunity may be lacking.
- The impacts that these two forms of wealth have on the natural environment should be considered, as well as how they differ in terms of their impact on the cultural commons of other cultures.
- How these two different forms of wealth influence the democratic process should also be discussed.
- As students acquire a more embodied understanding of the differences between the cultural commons and the industrial/consumer-oriented subculture, they need to consider how transforming of the cultural commons into commodities and monetized services affects the environmental commons.
- How to understand the connections between the intergenerational renewal of the cultural commons in ways that reduce the adverse impact on the environmental commons and the nature of ecological intelligence is important in itself. It also establishes the basis for considering a number

of misconceptions that are a threat to the local cultural commons and to the prospects of an ecologically sustainable future.

- Following a discussion of the nature of ecological intelligence, and how it will be expressed differently from culture to culture, there needs to be a discussion of the origins of the misconceptions that are reproduced in the meanings that most people associate with such words as "tradition," "individualism," "property," "progress," "environment," "freedom," "technology," "science," and so forth. The key question is: How have these misconceptions limited the development of ecological intelligence? The question of how different technologies, and the ideology that justifies their use, undermines the local cultural commons, as well as the diversity of the world's cultural commons, also needs to be considered. This should lead to examining how different technologies amplify certain ways of thinking, values, and relationships while reducing others. That is, can the mediating characteristics of different technologies become part of the process of cultural colonization?

- Consideration should be given to how the transformation of scientific discoveries into meta-narratives that explain the development of cultures, such as the theory of evolution which is now being extended to explain cultural memes, as well as the argument made by some scientists that they possess the only valid approach to knowledge, contribute to undermining the diversity of cultural commons—and, by extension, the environmental commons of the world. There also needs to be a discussion of the background knowledge students needs to possess in order to challenge the injustices that are part of some cultural commons. This would include a discussion of the background knowledge necessary for resisting various political and economic forces that are transforming the cultural and environmental commons into the private property of individuals and corporations.

- Invite students to consider whether the spread of ecological intelligence among the general population will be necessary if they are to have a sustainable future. Also have them consider whether ecological intelligence will lead to a radical change in how private property is understood. The changes that represent a shift away from the traditional idea of private ownership of property, ideas, and innovations also need to be discussed.

Two Suggestions for Integrating What is Learned in Schools with the Intergenerational Knowledge of the Cultural Commons

Public schools and universities need to provide leadership in connecting students to the wealth of the cultural commons. This is especially important today, as real wealth is not attained by depleting the wealth of the environmental commons—the hydrocarbons, oceans and streams, soil, forests, and minerals—in order to meet the public's consumer addiction. The first suggestion for exercising leadership is to establish a connection between the local high school and what can be called the community sustainability council. The council would consist of members of the community who possess knowledge of daily living practices that reduce dependence upon the money economy as well as have a smaller ecological footprint. The intergenerational knowledge and skills to be shared with the students through a combination of a class format and field experience would range from how to conserve water, plant eatable yards, reduce the use of electrical power, avoid the use of toxins, preserve (canning, in the old vernacular) fruits and vegetables, to preparing meals from local sources. As the knowledge and skills would be shared by members of the local community, it would reflect an understanding of the unique characteristics of the bioregion. For example, knowledge about how to increase the number of pollinators and diversity of birds, as well as the types of vegetables that thrive in different seasons and in different soils, would have practical benefits. On their own, students are not likely to learn the knowledge and skills accumulated by the long-term inhabitants of the region. And as the money economy continues to slide, along with how automation reduces the need for workers, the students will begin to recognize that greater dependence upon the knowledge and practices that sustain the local cultural commons is a way of escaping the debilitating impact of economically driven poverty.

A second proposal for how the local high school can take a leadership role in revitalizing the local cultural commons would be for students in the social studies class to maintain a website that enables members of the community to network with each other in meeting the following needs:

- Enable the unemployed and under-employed to contact various mentors in the community who are engaged in cultural commons activities—ranging from food security, creative arts, craft knowledge and skill, to volunteering, and developing social organizational skills. The first step would be for high school students to conduct a survey of the mentors in the community, as well as the different activities that are part of the local cultural commons. When the unemployed and under-employed are able to network with others in the community, they will be more likely to discover interests, talents, and the benefits of community participation that they did not have time for when they were caught in the cycle of working in order to consume, and to prevent a further slide into debt.
- Enabling members of different social groups to share their knowledge of how to prepare nutritions meals from locally available basic ingredients that can be obtained at a fraction of the cost of the processed foods handed out by food banks. This will empower people with the knowledge and skills necessary for meeting their nutritional needs with basic ingredients that ethnic groups have relied upon in the past. It will also provide a community alternative to the current practice of distributing packaged foods to the unemployed that contain many unhealthy chemicals.

- Enable farmers to communicate with others in the community about when their fields and orchards are open for gathering free vegetables and fruits. A computer network that connects local farmers with a community clearinghouse for those in need would be especially important, as well as ensuring that a manageable number of people visit these farms.

- Enable people who have already made the transition to voluntary simplicity, or have less need for an income connected with full time employment, to communicate their willingness to engage in job sharing. The network would enable people seeking part-time work to communicate with people willing to make the transition to part-time employment. There will be a number of issues, depending upon the nature of employment that will need to be worked out and agreed upon. The dominant issue, however, is to strengthen the sense of community by helping reduce the level of unemployment and hopelessness that will continue to be a problem as automation, downsizing, outsourcing, and economic systems continue to undergo change.

- Enable members of the community to barter with others who possess skills and can provide services, thus restoring the traditional understanding of the market as an exchange of goods and services that enhance the self-sufficiency of the local community.

- Enable individuals and groups needing some form of assistance to communicate with members of the community who are willing to volunteer their time and energy.

As is often observed, new opportunities emerge during life-altering crises. We are now facing the consequences of excessive consumption, the production of goods that far exceeds the needs of sensible people, and financial speculation driven by pure greed. The major disruptions caused by these excesses are occurring at a time when further automation is likely to leave many more people below the poverty line—or perilously close to it. We are also on the cusp of environmental changes that will create even greater challenges, as the scale of environmental change will lead to vast numbers of people here and abroad becoming environmental refugees, as the ecosystems they previously relied upon for their livelihood become too degraded to support even a subsistence lifestyle.

There are increasing references in both scientific journals and the media to the need to introduce changes that will slow the rate of environmental degradation. Unfortunately, most people still give only lip service to making changes, and the changes they do make are largely limited to recycling their trash into the proper disposal bins. Progress is being made in introducing new energy-efficient technologies and retrofitting buildings. Expressing concern about the environment, which for many is little more than giving expression to what is politically correct, is nevertheless a sign of an opening to learning about the important challenges that lie ahead. Too often, the inability to act on current understandings about changes in the Earth's natural systems is a result of an educational system that indoctrinated people with the ideas and values that are now failing us. The local cultural commons do not have to be created by government, nor is their existence dependent upon implementing the abstract theories of academics. They can be traced back to the earliest human societies, and they continue to exist even in the most oppressive circumstances.

Religious groups are now struggling to correct a myth of creation that represented, in one powerful account, that "man" was put here to name and subdue the natural world. Even real-estate professionals must now pass a test on the sustainable characteristics of the houses they are trying to sell. Ironically, their awareness that houses must now meet environmental codes is way ahead of the thinking of most public school teachers and university professors. Aside from the small number of environmental educators, and a minority of faculty in colleges and universities who are pushing the boundaries of their areas of inquiry in ways that address environmental issues, the vast majority of faculty who have the potential for influencing young minds, especially professors in colleges of education, seem unable to recognize that the modernizing paradigm they learned from their professors does not lead to understanding the solution. The emphasis on individualism and progress, along with the measurement and control technologies that still dominate the field of teacher education, continue to perpetuate the silences and prejudicial language that make the non-monetized and intergenerational-connected activities and relationships within communities appear as sites of backwardness.

The previous discussion of the political economy of the cultural commons is intended to address some of the silences that still contribute to teacher educators thinking that the ecological crisis is being met by scientists, technologists, and environmental educators who are, in many instances, limited in their understanding of the cultural roots of the ecological crisis. While learning how to foster the ecological intelligence of students will be a major challenge, especially since the practice of ecological intelligence requires abandoning many Enlightenment assumptions, encouraging students to learn from the people who are sustaining the wealth of the local cultural commons should be much easier—particularly when it involves face-to-face relationships and mentoring in activities that fosters the students' self-discovery of community-centered interests and talents.

Nothing new needs to be invented and promoted. Rather, the role of public schools and universities in revitalizing the local cultural commons requires putting aside certain misconceptions inherited from earlier thinkers who were addressing an entirely different set of problems, and giving attention to the local practices that have not been monetized—and that have a smaller adverse impact on the environment. Auto-ethnographies, the importance of face-to-face intergenerationally connected communication, and a greater sensitivity to the kinds of experiences that enable students to discover talents, as well as who they are as members of a community, is the way forward. And if teacher educators, and professors in the other areas of educational studies, can make this turn, perhaps they will then help students obtain a different understanding of wealth—one that takes account of what is shared with others and is personally fulfilling in ways that differ from owning what has been industrially produced for a mass market. Whether faculty in the social sciences and humanities begin to address the cultural roots of the economic and ecological crises, and the ways they have been complicit in the globalization of the industrial/consumer-oriented culture, is still problematic.

Reference

Batalla, G. B. 1996. *Mexico Profundo: Reclaiming a Civilization*. Austin, Texas: University of Texas Press.

Daly, H. E. 1991. "Consumption: Value Added, Physical Transformation, and Welfare." In *Getting Down to Earth: Practical Applications of Ecological Economics*, edited by Robert Costanza, Olman Segura, and Juan Martinez-Alier. Washington D. C.: Island Press, 49–59.

Shiva, V. 1993. *Monocultures of the Mind*. London: Zed Books.

8 The Cultural Mediating Role of the Professor— Across the Disciplines

Many professors take pride in the fact that they have never taken a course in pedagogy. In most instances, they lack an understanding of what is learned in such courses, which are usually located in a college of education. And in this case, ignorance is bliss as most courses on pedagogy are based on the double bind thinking discussed in the previous chapter. That is, most such courses reinforce the idea that language is a conduit, that change ("emancipation" is the teacher educator's code word) should be constantly promoted, that students should learn to construct their own knowledge, that technology is both culturally neutral yet essential to participating in the global economy, that teachers should help students understand the issues of race, class, and gender—thus contributing to the ability of marginalized groups to take their place as equals in a consumer-oriented society, and that traditions are impediments to progress. While some of these concerns are valid, most courses that address how teachers should understand their pedagogical responsibilities reproduce the silences and misconceptions that have been a hallmark of the field for generations.

Professors may be correct in their judgments about courses in pedagogy, but they are incorrect in thinking that successful teaching and learning is simply dependent upon well-thought-out and well-presented lectures, reliance on the Socratic method, an increased use of computers and Power-Point presentations, and on the use of smaller discussion groups. There are some fundamental characteristics of learning that professors, regardless of their discipline, need to understand—including how to take these characteristics into account when mediating between what the students bring to the teacher/student relationship and the new understandings that the professor hopes to introduce. Most of these characteristics were discussed in earlier chapters, but they need to be reiterated in order to clarify more fully what is meant by referring to the professor's role as that of a mediator.

A universal characteristic of teaching/learning relationships is that both the student and professor take for granted a large body of beliefs and culturally specific assumptions that influence how new ways of understanding are presented and learned. That is, the largely taken for granted interpretative frameworks of both the professor and student will influence what is heard, seen, and how it is understood. In other words, the professor and student do not stand in a relationship involving autonomous individuals. Rather, they represent culturally and biographically distinct traditions of thinking and culturally mediated embodied experiences. What the professor takes for granted will frame both the language that is used to communicate with the student, as well as the silences that are due to her/his restrictive vocabulary. And if the student is encountering something that has not been learned before, she/he will not be aware of how the silences in the professor's course undermine the ability to address future problems.

Another characteristic of the professor/student relationship, regardless of whether it is in the sciences, social sciences, humanities, or one of the professional programs, is that few professors and even fewer students will be aware that the conduit view of language that is such a common feature of most classroom discourse, including what is read in books and on the computer screen, reproduces the metaphorically layered patterns of thinking that, as Nietzsche put it, fit "new material into old schemas." How the conduit view of language contributes to misunderstanding, like the ever-present fog of taken for granted beliefs and assumptions, is a constant of which professors need to be aware. It takes a special effort to become aware of what the student takes for granted, and it is even more difficult for the professor to become explicitly aware of her/his own taken for granted cultural assumptions.

Other characteristics of the languaging processes that are at the center of teaching and learning include the likelihood of ethnocentric thinking and the combination of silences and prejudices that have relegated the intergenerational knowledge that sustains the local cultural commons to low status. As pointed out earlier, both professor and student too often frame what is being learned as the expression of progress, and to view non-western traditions and knowledge as less advanced—and thus not worthy of learning from. Most of the root metaphors (meta-cognitive schemata) discussed earlier come into play as the student is socialized to think within the professor's discipline. There will be times when these root metaphors contribute to new understandings, and many more times when they perpetuate the cultural patterns that are major contributors to the current acceleration of the environmental crises. Context, tacit understandings, cultural differences, and now what contributes to an ecologically sustainable future and what undermines this possibility, are all considerations that need to be taken into account by the professor.

Perhaps the most important challenge professors face is recognizing how their own socialization within their discipline may continue to perpetuate the deep cultural assumptions that earlier generations took for granted. These assumptions were passed on as part of their own graduate studies when their own professors were unaware that the assumptions were constituted before

there was awareness of environmental limits—and before there was an awareness of the smaller adverse environmental effect of most activities that are part
of the cultural commons. As mentioned before, the prior socialization of the
professor may be so out of touch with today's rapidly changing environmental realities that she/he may not be able to recognize the evidence and implications of ecological collapse. There are accounts of when the indigenous
people first encountered the tall-mast ships of the European adventurers in
the local harbor, they initially had no way of understanding what they were.
Nor did they have any way of knowing how the arrival of these ships would
change their world. This may be the same problem that is faced by professors who continue to take for granted the assumptions that are still shared
by the cohort of colleagues who were mentored by professors who did their
graduate work decades ago.

Before suggesting that professors need to take on an additional responsibility that will make their task even more complicated, I want to identify another
problem that needs to be recognized, even if it remains intractable. It's the
problem that Carl Schmitt used as justification for an authoritarian system of
government that classified liberals as the enemy of the state. His basic argument was that when the state faces a serious external threat (he was referring
to Nazi Germany's need for national unity in supporting its planned wars of
aggression) the liberals became an internal threat because of their tendency
to argue over every issue, to pursue their own agendas, and to be unable to
agree on what constitutes a common external threat (1976). Schmitt's arguments are now being illustrated by today's market liberal politicians who
view the terrorism stirred up by their agenda of economic globalization as a
threat to the stability of the United States. However, Schmitt's point is more
relevant to what is now a genuine external threat: the rate and scope of the
ecological crises.

Given the double bind thinking discussed in the previous chapter where
one of the analogs for understanding what it means to be a liberal thinker
is that of being an autonomous individual who is guided by her/his "own"
powers of critical rationality, the question now becomes one of whether
social justice and market liberal professors can reach a consensus that global
warming, and the many changes occurring in other ecosystems, are a
genuine threat. Will they be able to reach consensus that this should be the
main priority in undertaking major university reforms—and by, extension,
reform of public schools? In suggesting the role that professors and public
school teachers needs to play as mediators who help students understand
the differences between the local cultural commons and the industrial/
consumer-oriented culture they daily move between, I will make the optimistic assumption that as generations in the past abandoned the idea that the
earth was flat, and later recognized that it was not the center of the universe,
this generation of market and social justice liberal professors will make a
similar adjustment to what now constitutes the new scientific evidence of

an even more fundamental change that has not occurred since the last great mass extinction.

Role of a Mediator and Why is it Important in Terms of Addressing the Cultural Roots of the Ecological Crises

One of the characteristics of participants in many local cultural commons activities, such as maintaining community gardens, mentoring relationships, promoting social justice issues at the local and national level, is that they involve democratic decision making. In order to participate in this process, it is necessary that the members of the cultural commons possess communicative competence. That is, if individuals lack the language necessary for naming the cultural commons activities that are being threatened by different forms of enclosure, and if they do not understand how the aspect of the cultural commons being threatened contributes to the well-being of the larger ecology of community/environmental relationships, they will be limited in their ability to resist the external forces of enclosure. Similarly, if they cannot name other traditions of the cultural commons, such as the language, narratives, laws, and other taken for granted patterns of interpersonal relationships that support traditions of discrimination, they will be unable to engage in the democratic process of bringing about needed reforms. Examples of this process can be seen in how the feminist, civil rights, and migrant worker movements, among others, demonstrated the importance of being able to name the nature and sources of prejudice and discrimination as the first step in breaking the hold of taken for granted beliefs and practices—which often bind both the exploiter and the exploited to traditions that have not been made explicit and challenged.

In the context of global warming, communicative competence involves more than what is too often modeled by the elaborated speech code of the professor who can justify her/his assertions, who can cite evidence based on research, and who can talk endlessly without ever acknowledging that human activity is responsible for the changes now occurring in the oceans, atmosphere, and animal and human habitats. Communicative competence that is not based on double bind thinking requires that the student and other members of the cultural commons possess an explicit knowledge of what is being threatened by market forces, new technologies, and various forms of religious fundamentalism. They also need to possess knowledge of the forces behind various forms of enclosure, such as the ideology and corporate agenda behind the current efforts to enclose (that is, take away) long standing traditions of civil liberties, the gains in the labor movement, the face-to-face traditions of mentoring (which are now being replaced by DVDs), the intergenerational traditions of children's play, and so forth.

If the taken for granted experiences in the cultural commons, as well as in the market-oriented culture, do not become part of the students' explicit knowledge and vocabulary, they will be unable to recognize the reasons for the downward spiral into a state of dependency and poverty that many people are now experiencing. This condition of poverty includes more than the lack of economic resources; it also includes the deprivation of symbolic culture that is the source of meaning, expressive arts, patterns of moral reciprocity, and narratives of how to live lightly on the land and in mutually supportive ways. To make the point more directly, if students cannot name what is being enclosed, they will be unable to resist the forces of enclosure. In short, they will be unable to change what is a destructive tradition. A few examples may be helpful. The failure of professors in the past to help students recognize and thus make explicit the many expressions of gender bias that were a taken for granted part of institutions, the legal system, the workplace, and other areas of the cultural commons, limited the student's ability to acquire the language necessary for exercising the communicative competence in bringing about fundamental social changes. It was the feminists who named the patterns that provided others with the language necessary for politicizing what previously was part of people's taken for granted reality. The feminists were doing "thick description," while those who were perpetuating the patriarchal patterns at all levels of social life were defending the ancient analogs they associated with the word "woman."

Similarly, a lack of knowledge of the characteristics of fascist societies, which came to power in Europe between the two World Wars as a result of the rise in social chaos and a desire for a powerful centralized authority, will leave students without an explicit awareness of how the technologies of total surveillance and the combination of the market liberal and corporate agenda of economic globalization may be putting us on the same slippery slope. Before taking up the issue of how the professor's mediating role can address the critical problems of double bind thinking, as well as contribute to the students' communicative competence in making decisions that are ecologically viable and assist in initiating a fundamental re-ordering of the priorities of the university, it is necessary to again identify the Achilles' heel of higher education. That is, the patterns of thinking most professors take for granted make it difficult for them to reflect on how they may be major contributors to the industrial/consumer oriented culture that is partly responsible for the billions of metric tons of carbon dioxide that are changing the chemistry of the atmosphere and the world's oceans.

I suspect that as scientists document the human impact that is causing the melting of glaciers, the changes in habitats to which plant and animal species are unable to adapt, and the dislocation of huge numbers of people as crops fail and potable water becomes in even shorter supply, more professors will include environmental issues in their courses. And as the nature of the environmental crisis becomes obvious to the point that denial is no longer possible, some may even begin to make explicit and to question how the deep cultural assumptions being reinforced in their courses are preventing them from taking seriously the only proven alternatives to the current reliance on hyper-consumerism to which many Americans are now addicted.

The alternatives are the cultural and environmental commons that humans have relied upon since their earliest beginnings, and that have been relegated to low status by universities. As these changes occur, professors will need to take seriously their role as mediators.

Mediating involves helping students become explicitly aware of the differences between their experiences in the local cultural commons and their experiences in the sub-culture characterized by the cycle of working for money in order to purchase what too often will quickly be replaced by new consumer products, falling further behind in credit card debt, becoming increasingly stressed—thus becoming more dependent upon the pharmaceutical industry, and then facing retirement without either an adequate economic source of support or a knowledge of how to participate in the local cultural commons. This may be an over simplified account of the cycle of life in the industrial/consumer culture. On the other hand, it is accurate in terms of the greed of the power elites who grant themselves multimillion dollar severance packages while their corporations outsource work to countries with lower wages, and even lay off the higher paid workers in order to replace them with workers who are paid the minimum wage. There is little left of the moral reciprocity that the labor movement and social justice groups forced on corporate America. What remains of moral reciprocity exists mostly within various intergenerationally connected groups carrying forward different traditions of the cultural commons, including the traditions of mentoring in various skills and in volunteerism.

Before discussing what is involved in the professor's (and classroom teacher's) role as a mediator, it is necessary to address a possible misinterpretation. Because I sometimes refer to the cultural commons of indigenous cultures, critics often claim that while my ideas possibly have relevance to rural America they have little relevance for suburban and urban America. These critics obviously do not understand that the local cultural commons may be expressed differently in rural and in urban America—and that language, civil liberties, intergenerational knowledge, skills, ceremonies, narratives, mentoring relationships, and so forth, are also part of the cultural commons in urban areas. That is, the tension between what is not dependent upon monetized relationships and the forces of economic enclosure exists in *all communities.*

Another misinterpretation that critics impose on my proposals for reforming universities is that I am suggesting that the entire curriculum focus on the tensions between the commons and market forces—and that, by extension, I am suggesting that all forms of scholarship not focused on these tensions should be abandoned. This is definitely not what I am proposing. I know that the long-held interpretative frameworks that underlie current scholarly and teaching interests will continue to prevail—partly because they are taken for granted and partly because many of these traditional scholarly interests are important for other reasons. I am a realist in recognizing that even when new ideas are supported by evidence of extreme importance, such as global warming, the influence of past ways of thinking will only yield slowly to thinking within this new (actually old) paradigm that places community over the importance of what can be manufactured and sold for a profit. Also recognized is that many of the achievements of the past will be integrated into this new paradigm. In my most optimistic moments I am hoping that professors will *begin* to address the tensions and misconceptions that will contribute to revitalizing the cultural commons, to democracy at the local level, and to making the transition to fostering ecological intelligence.

Mediating as Making Explicit

At every age level, students are involved in experiences in the cultural commons and in the consumer/industrial culture. They generally move between them without being explicitly aware of what the differences are in social relationships, language, dependencies, forms of empowerment, influence on skill development, and so forth. Some of these experiences involve different forms of enclosure, such as undermining self-confidence, marginalizing the exploration of personal talents and skills, fostering greater dependence upon monetized relationships, eliminating privacy, replacing the practice of moral reciprocity with the pursuit of self-interest and material wealth, and so forth.

Preschool children move between the oral traditions of the family (not all of which are supportive of social justice and good environmental citizenship) and the computer-based entertainment and communication that reinforces the industrial/consumer-oriented mind-set. In the early grades students are involved in oral and print-based forms of thinking and communicating. They also participate in various creative arts that are only marginally dependent upon consumerism, and the various arts promoted by the entertainment industry.

At the university level, the language framed by the deep cultural assumptions that support the further expansion of the industrial/consumer culture becomes a more prominent part of the students' experience. This is accompanied by the silences and prejudices that further marginalize an awareness of the cultural commons that are also part of their everyday life. Few university students, for example, can explain the nature of metaphorical thinking. Nor are they knowledgeable of the history of root metaphors and how the vocabulary they rely upon on a daily basis is largely dictated by these root metaphors. And few students are able to recognize when scientists are straying into the quagmire of scientism. In addition, the different technologies that teachers and university professors encourage students to rely upon make the need for doing a thick description of the student's embodied/cultural experiences appear totally irrelevant.

Mediating thus starts with the students' description of what they are experiencing as they participate in different aspects of the cultural commons and the consumer-oriented culture. The thick description is different from the form of learning in which the classroom teacher, professor, and software program start with telling the student how to think about different aspect of their everyday world—or history, or other cultures. What the anthropologist, Clifford Geertz referred to as thick description involves naming different relationships, feelings, sense of empowerment, discovery of interests, awareness of what cannot be communicated, how the activity or object dictates how one should act and think, and so forth. Thick description enables the students to connect an activity or relationship within the larger network of relationships that Bateson referred to as the cultural and natural system pathways through which information is passed. Making explicit the ongoing exchange of information that circulates through the interconnected ecologies of culture and natural systems acts on the actions of others, including the student's experience. In effect, this process of doing thick description can be understood as doing a personal ethnography.

Becoming aware of one's own cultural patterns is difficult, and the taken for granted nature of these patterns adds to the difficulty. And this is where the mediator role becomes especially important. As we can see in so many instances of taken for granted beliefs and practices—ranging from the person who takes for granted that most of life's needs can be met through consumerism, the person who takes for granted that women or some other group are inferior performing certain tasks, and the chemist working for Monsanto who takes for granted the cutting-edge and progressive nature of her/his research—there is little that appears problematic and thus there is little need for reflecting on what is hidden behind the otherwise taken for granted experience. Thought becomes focused on other tasks at hand. The mediator's role becomes clearer when we consider how students (as well as adults) move seamlessly from cultural commons experiences to consumer/monetized experiences. As the taken for granted nature of this transition marginalizes awareness of how both areas of experience involve multiple consequences in terms of personal development, the impact on others, and the natural environment, there is a need for a mediator. The responsibility of the mediator, however, is not to explain what the students' choices should be or what judgment they should make. Rather, the mediator's role is to ask questions about the differences in experience that will enable students to become explicitly aware of what otherwise is taken for granted. This increased state of awareness leads in turn to developing communicative competence in being able to describe how different cultural practices and relationships amplify and reduce personal abilities, influence relationships, and impact the environment.

The people who listened, asked questions, and generally supported the efforts of women to give voice to what otherwise had been taken for granted, were modeling the role of the mediator. In helping students become aware of the differences between the spoken word and technologically mediated communication, between developing a musical skill or participating in a family meal and downloading a musical performance and eating at a fast food outlet, between engaging in a craft such as weaving and working on an assembly line, students are likely to become more aware of different aspects of the cultural commons and what is being lost when they are enclosed by the forces of technology and the market. Perhaps one of the more important insights is in learning the range of cultural commons activities that lead to the discovery of personal interests and the development of talents—and the awareness that the cultural commons enables one to live a meaningful life that is less dependent upon participating in a money economy and that is less environmentally destructive.

As the mediator asks the questions that have not occurred to the students, and brings into focus relationships and interdependencies as well as the modern forces of enclosure, the student is then developing the vocabulary and explanatory frameworks essential for making decisions about the differences between commons and market-oriented activities. This approach to mediating—that is, helping the student become explicitly aware of the network of past and current relationships that would otherwise be taken for granted—reinforces the idea that the student's embodied cultural experience is important to learn about, and to develop the communicative competence necessary for articulating what needs to be conserved and what needs to be radically changed. The role of the mediator is, in effect, critical to revitalizing the possibilities of local democracy. If students are unaware of the importance of the local cultural and environmental commons, they are likely to accept the privatizing and monetizing of different aspects of the cultural commons as the latest expressions of progress. This acceptance of the growing dependence upon outside technological and economic forces that are under the control of corporations and market liberal and libertarian ideologues is one of the most prominent features of today's world.

It is relatively easy to identify the characteristics of the cultural commons, as well as the different forms of enclosure. What is more difficult to address is how to ensure that the classroom teacher and professor are aware of the complexity of the local cultural commons, and how important they are to living less consumer-dependent and environmentally destructive lives. I have had many experiences with highly educated people who have responded to the mention of the cultural commons by arguing that we cannot go back to a pre-modern stage of existence. Further conversations revealed that they engaged in various cultural commons activities at a very high level of competence. The problem was that they had not thought of their activities as part of the cultural commons, and thus had not thought about the importance of educating students about the community-centered alternatives to an economic system that relies more on automation than wage earners—and that is ecologically destructive. More typical are the classroom teachers and professors who move seamlessly between the two sub-cultures, and are unaware of the questions that need to be asked—about their own taken for granted experiences and what their responsibilities are toward the next generation that is becoming even more addicted to technologically mediated communication and awareness. Clearly, this is an area of university reform that needs to be addressed. Although many professors are not explicitly aware of the complexity of issues related to the enclosure of the world's diversity of cultural commons, they are nevertheless able to make an important contribution—assuming the issues are raised.

Mediating as Introducing a Knowledge of the History of the Commons and the Forces of Enclosure

Engaging students in a comparative examination of their experiences in the cultural commons and in the consumer-dependent relationships needs to be framed in terms of assessing which contributes to a smaller ecological footprint and to a more socially just and self-reliant community. But it needs to go beyond the thick description that makes explicit the cultural and environmental context—including what forms of relationships and patterns of thinking are being reinforced and how these patterns vary in terms of different cultural assumptions. That is, the classroom teacher and professor need to bring an historical perspective to the comparative analysis. For example, students may be involved in community experiences that are centered on one or several performing arts, and they may also be involved in downloading commercially produced art onto their iPods or onto a new technologically dictated format that cannot even be imagined at this time. The students need to examine the history of cultural developments that contributed, for example, to the transition from the various arts from being integral to the community's ceremonies, from being storehouses of knowledge of moral relationships, from being ways of transforming (as Ellen Dissanayake points out) the mundane aspects of everyday life into a realm of experience that is special and transcendent, to being what is valued because the market has designated it is a source of profits—and because it is produced by a person whom the market has elevated to celebrity status.

Other comparisons in which the historical perspective needs to be introduced include learning about the cultural developments that subordinated craft knowledge and skill to the need to find more efficient and low-cost methods of production, and now to replace workers entirely with computer-driven systems of production. For example, students should know what led to giving high status to print-based knowledge and to representing orality as an unreliable source of knowledge. The tensions between civil liberties that are protected by the tradition of separation of church and state, and the political/religious forces that are working to undermine this separation in order to create a theocracy, also need to be understood in terms of the history of the religious wars that ravaged Europe for hundreds of years.

If students are going to become aware of how the Orwellian use of language, which has become such a prominent feature of today's political discourse, has marginalized an awareness of how all the participants in cultural and environmental ecologies are interconnected, they will need to examine the history of the layered nature of metaphorical thinking. Every aspect of the cultural and environmental commons, as well as the ideological and technological forces of enclosure, has a history. Students also need to learn about the history of the enclosure of socially unjust traditions that have been a prominent feature of some cultural commons. The students' communicative competence is as much dependent upon a knowledge of the history of the development of the cultural commons as it is on knowledge of the history of the forces that are contributing to the current processes of enclosure. Again, it must be stressed that this historical perspective enables students to recognize the misconceptions of the past that still dominate current thinking and policies, as well as to recognize the traditions that grew out of past struggles that need to be carried forward and intergenerationally renewed.

Why a Department of Cultural Commons Studies is Needed

When we compare the nature and rate of environmental degradation, which range from the changes in weather patterns that are melting glaciers that are the source of water for millions of people and are causing droughts that make huge areas uninhabitable, as well as the collapse of edible fish stocks that are an important source of protein for an expanding population, the suggestion that university reforms should include the creation of a department that has the responsibility for providing courses that introduce students to an in-depth knowledge of the cultural commons and the forces of enclosure may appear as too little and too late. Nevertheless, this proposal needs to be viewed in the light of what is problematic about the current approaches to introducing environmental issues into courses where the conceptual framework too often is dictated by the traditional assumptions upon which the discipline has been based. Except for the sciences, the efforts to introduce environmental issues into courses in the social sciences, humanities, and professional programs represent introductory efforts. But their introductory nature is only part of their shortcomings.

As mentioned earlier, courses in sociology, history, philosophy, literature, economics, political science, education, law, and so forth, follow the general pattern of more traditional courses that are thought to be strengthened when students are introduced to a wide range and often unconnected series of readings. Aside from the fact that few, if any, of these courses introduce students to writings on the cultural commons and to doing a thick description of their experiences in the local cultural commons, there is another major shortcoming. A course that introduces students to the writings of Henry David Thoreau, John Muir, Aldo Leopold, Rachel Carson, Gary Snyder, Wendell Berry, and Vandana Shiva is a valuable learning experience for students, in that it introduces them to profoundly different ways of thinking about human/environmental relationships, and to what gives a deeper sense of meaning to life.

To cite another example that appears to have similar strengths in terms of awakening students from the industrial culture's dream of a life of ever-expanding prosperity (and if that fails, the prospects of life in an inter-planetary settlement) is an introductory course in environmental studies that introduces students to short excerpts from the writings of David Abram, Stephanie Mills, Arne Naess, Lao Tzu, Rene Descartes, a Cherokee creation story, and other environmental writers. Courses that follow this approach have a major shortcoming that can only be corrected by having a depart-ment that has as its central focus the study of the cultural commons and the myriad forms of enclosure. The problem is that students who take these survey courses, as important as they are, will graduate without an in-depth knowledge of the different ways in which cultures renew their cultural com-mons, including how some cultures have learned to live within the sustaining capacity of the natural systems in their bioregion.

A department of cultural commons studies would have the advantage of not having the environmental issues determined by faculty whose main area of intellectual competence is based on their past training in a traditional dis-cipline. Rather, the faculty in this department would be better able to ensure that students encounter the conceptual framework that introduces them to the nature and diversity of the cultural and environmental commons, as well as the different forces of enclosure. This conceptual framework is needed to understand the importance of how the insights, silences, and prejudices of a wide range of environmental thinkers were influenced by the cultural assumptions of their times, why their ideas were widely ignored by the larger society, and their relevance for understanding the tensions between the local cultural commons and today's forces of enclosure. Unless students acquire this conceptual framework, they will be less prepared to recognize a wide range of cultural issues that will likely not be discussed in environmentally oriented courses taught in the traditional disciplines, and the survey-type courses that introduce students to the writings of major environmental thinkers.

The basic course offered by the department needs to introduce students to the role that the languaging processes of a culture play in constructing what will be taken as the common-sense daily reality—what can also be called the taken for granted storehouse of cultural knowledge and values. Other needed basic understandings include the layered nature of the metaphorical language that is taken for granted by everyone, even by people on the cutting edge of their field of inquiry—as well as how this language reproduces the miscon-ceptions of earlier times. The nature of the historically rooted misconceptions needs to be a central focus in this introductory course. Students also need to understand the differences in cultural approaches to storing and renewing intergenerational knowledge, such as the differences between oral and literacy traditions (which also vary from culture to culture). The British and Marxist anthropologist, Jack Goody, argues in *The Domestication of the Savage Mind* (1987) that the divide between oral and literate cultures is far more signifi-cant than the divide between social classes. The question that is not likely to be asked by students reading and discussing a wide range of environmental writers, and early philosophers who got it completely wrong, relates to how the tradition of literacy (and now the increasing reliance on computer-mediated thinking and communication) contributes to the inability of students to hold the users of language accountable in terms of accurately representing local contexts, tacit and embodied experiences, and an awareness of how current cultural practices will impact future generations. I suspect that there are few courses that introduce students to the wisdom of environmental writers that also engage students in a discussion of the different impacts that computers have on the cultural and environmental commons—impacts that are both positive and destructive.

This basic course should also introduce students to the thinking of Gregory Bateson's understanding of how human intelligence is encoded ("immanent") in the material culture that, in turn, influences the natural systems in which the material culture is embedded. There is a tendency on the part of university graduates to view the various expressions of material culture as things, distinct objects, buildings, and so forth, and to lose sight of the fact that they embody an earlier form of intelligence that may have been based on cultural assumptions about an anthropocentric, mechanistic, and inherently progressive world. A deep knowledge of Bateson's ideas about intelligence, including how intelligence is a mix of the culture's recursive conceptual patterns and the individual's ongoing participation in the multiple pathways through which different participants in the larger ecology communicate differences, is essential to overcoming the misconception that thinking is an activity occurring in the brain of the individual. Students who take subsequent courses that address different environmental and cultural commons issues need to reject this increasingly reductionist view of intelligence if they are going to learn from the thick descriptions of the differences between their experiences as they move between the monetized and non-monetized activities and relationships within their communities.

There are several other reasons for establishing a department of cultural commons studies. The first relates to the need for members of the department to recognize when courses could be strengthened by involving faculty from other disciplines who can introduce the unique insights of their disciplines into the discussion of various aspects of the cultural commons and the different forms of enclosure. What is being suggested here is a reversal of the current approach where faculty, who have little background knowledge about the nature of the cultural commons and the different forms of enclosure, introduce environmental issues into courses that are still dominated by the deep assumptions of their discipline. In these situations, the environmental issues are more peripheral to the deep cultural assumptions that the professor in philosophy, economics, sociology, literature, and so forth, may take for granted. This add-on approach, which leaves unexamined the deep cultural assumption underlying the professor's primary area of interest and scholarship, may have a long-term influence on the students' way of thinking (even after they have forgotten what they learned from the environmental writers).

Faculty in the department of cultural commons studies would, in effect, have the responsibility for knowing which faculty in other departments could make an important contribution. If I taught a course that focused on how western philosophers contributed to the prejudices and silences about the importance of the cultural and environmental commons, rather than writing about it myself, the course would be strengthened by involving faculty from the department of philosophy who could provide insights about how the ideas of different philosophers validated the argument that part of our current patterns of double bind thinking can be traced back to the acclaimed giants of western philosophy. They may have been able to provide counter-evidence; that is, if they are aware of the cultural commons, differences in cultural ways of knowing, the ways in which abstract theory undermines the development of ecological intelligence, and the environmental issues that the major western philosophers relegated to the silences that support the state of cultural amnesia.

The second reason for a separate department that goes beyond that of ensuring that the courses are grounded in a deep knowledge of why so many cultural patterns are taken for granted, and an equally deep knowledge of the diversity of the cultural commons—including the many forms of enclosure—is that the faculty in this department should take on the task of organizing workshops for other faculty who become interested in eco-justice issues and the need to renew various aspects of the cultural commons. Faculty in this new department can also play a key role in coordinating seminars and conferences that frame the discussion of alternatives to the consumer-dependent pathway of economic globalization. Introducing students to the political and economic forces that have enclosed earlier expressions of the cultural and environmental commons—ranging from the community centered intergenerational practices relating to food, healing, creative arts, social justice (and injustice) practices—should be the responsibility of faculty from a variety of disciplines. However, it is most likely to be undertaken if the faculty in a department of cultural commons studies serve as a catalyst for introducing this new area of understanding. This cultural commons pathway to a post-industrial future is still being treated as low-status, and the intergenerational knowledge upon which it is based is still considered as irrelevant to what university students should be learning—even in the environmental courses that are attracting more and more students.

The proposal that a separate department should be established is unlikely to be taken seriously unless faculty are able to recognize yet another conceptual double bind: namely, the need for faculty to overcome the silences and prejudices in their own education that may lead them to ignore this proposal as unworthy of their attention. As most faculty still consider the environmental crises as unrelated to their scholarly interests, the effort to overcome this current addiction to self-deception is going to require the dedicated and persistent effort of a few faculty. As changing human consciousness is exceedingly complicated and slow beyond what should be expected of rational people, we need to keep in mind that the most fundamental changes in recent times, such as those introduced by early feminists, Mahatma Gandhi, Rachel Carson, Martin Luther King, Jr., and Aldo Leopold, were started by a small minority who refused to go along with the taken for granted thinking and values of the times.

References

Bateson, Gregory 1972. *Steps to an Ecology of Mind.* New York: Ballantine Books.

Descartes, René. 2006. *Discourse on the Method of Rightly Conducting Reason and Seeking Truth in the Sciences.* New York: Oxford University Press.

Geertz, Clifford. 1973. *Interpretations of Cultures.* New York: Basic Books.

Goody, Jack. 1977. *The Domestication of the Savage Mind.* New York: Cambridge University Press.

Nietzsche, Friedrich. 1961. *The Will to Power.* New York: Random House.

Schmitt, Carl. 1976. *The Concept of the Political.* New Brunswick, N.J.: Rutgers University Press.

9 Educational Reforms in an Era of Global Warming and Digital Insecurities

One of the dominant challenges facing educational reformers educated in the last decades of the 20th century is recognizing how the conceptual frameworks for understanding the social justice issues of that era failed to address what scientists were reporting about climate change. Now that the rate of changes in the Earth's ecosystems is impacting people's daily lives in terms of droughts, changes in both the warming and acidification levels of the world's oceans, rising sea levels due to the melting of glaciers, and the disappearance of habitats and species, the emancipatory vocabulary handed down from the long tradition of social justice struggles in the West needs to be revised.

This does not mean giving up on educational reforms, or on challenging how the West's consumer-dependent industrial and now digital revolution continues the old forms of injustice and even new ones as the process of globalization continues. What needs to be revised is how the vocabulary that supports the West's way of interpreting progress has framed both the social justice agenda of educational reformers as well as the market liberal agenda of computer scientists, corporations, and the government's foreign policies. The vocabularies that continue to support the idea of progress that was part of the legacy of the Enlightenment thinkers of the late 16th and 17th century include **individualism** (with the ideal being the autonomous thinking individual); **change** and **innovation**; **critical inquiry** and **science** that overturn traditions; **transformative thinking**, **freedom**, and **literacy** that leads supposedly to objective knowledge and data that are the sources of **individual empowerment**, the **students' construction of their own knowledge and values**. But the most powerful word for legitimating ideas, policies, innovations, and the continual quest for the new and experimental is **progress**, which is understood as a linear move into the future that overcomes the backwardness of the past—that is, traditions. It did not matter that neither the Enlightenment thinkers, nor today's scientists such as Carl Sagan who claimed that "we give our highest rewards to those who convincingly disprove established beliefs" (Sagan, 1997, p. 35) understood that social justice achievements such as habeas corpus became a tradition. By reducing traditions to abstractions, Sagan and other anti-tradition thinkers avoided making explicit the taken-for-granted traditions they relied upon daily.

Following in the ethnographically uninformed thinking of John Locke, René Descartes, John Dewey, and Paulo Freire, most of today's critical pedagogy reformers continue to share the same Enlightenment view of traditions as sources of oppression and backwardness. What has gone largely unnoticed by these educational reformers and their many followers is that today's computer scientists and market liberal/libertarian heads of corporations also rely upon this same Enlightenment vocabulary to justify overturning cultural traditions throughout the world in order promote consumer-dependent and environmentally destructive lifestyles.

The irony is that none of the Enlightenment thinkers had a deep cultural understanding of the traditions they took for granted—even as they relied upon the many traditions built up over generations that enabled them to write their books. In effect, they relied upon the long-standing tradition of the early Greeks of encoding their ideas in the printed word that fosters abstract thinking that, in turn, marginalizes awareness of the lived cultural patterns that connect within different face-to-face relationships (Havelock, 1982). The anti-tradition abstract theorists of the past, as well as those still under the spell of the Enlightenment legacy that has morphed into today's progressive ideologies, ignored the fact that the crafts and skills used to create their dwellings, grow their food, and provide the artistic performances of the day were passed on from generation to generation, and even the early gains in social justice ignored how intergenerational traditions were passed forward. An even greater loss is that if these Enlightenment thinkers had been less ethnocentric and less ignorant of environmental limits, they might have learned from indigenous cultures the interconnections between their traditions and their development of ecological intelligence.

How the thinking of both John Dewey and Paulo Freire continues to reproduce the core Enlightenment assumptions that divide the world between the backwardness and non-intelligence of traditions and the progressive nature of critical/scientific inquiry can be seen in how Dewey explains the passive and outdated intelligence associated with traditions, as well as his account of the lack of intelligence of the "savages" (1916, pp. 58, 88, 396). Keep in mind that Dewey lived close to the Iroquois cultures, and during the time of the genocide of indigenous cultures that was occurring across the land, he remained silent. He also remained silent about the destruction of natural systems such as the millions of bison, the passenger pigeons, and the clear cutting of the nation's forests. Freire's Social Darwinian and Enlightenment-influenced explanation of the differences between three modes of consciousness is especially noteworthy. People living in the interior of Brazil, which he refers to as living a "semi-intransitivity" form of consciousness, "cannot apprehend problems situated outside their sphere of biological necessity." The second level of consciousness, which Freire labels "naïve transitivity" consciousness, must be overcome before reaching the most evolved level of "critically transitive consciousness" that involves critical awareness, democratic values, and the ability to continually transform the world (Freire, 1973, pp. 17–18). Both sets of tradition-oriented indigenous cultures identified by our two most acclaimed progressive thinkers, the indigenous cultures of Dewey's day that he refers to as "savages," as well as the indigenous cultures inhabiting the interior of Brazil, survived by developing ecological intelligence that enabled them to align their cultural ecologies with what could be sustained by the local natural ecologies.

In addressing how educational reformers can avoid the limitations in how the silences and misconceptions of Enlightenment thinking are still being carried forward by computer scientists, by scientists driven by hubris to genetically re-engineer the biological world, and by academics, teacher educators, and curriculum theorists, it is important to remember that one of the traditions of Enlightenment philosophers, and today's followers, was to ignore the life-ending experiences of cultures that failed to recognize environmental limits. In *Collapse: How Societies Choose to Fail or Succeed* (2005), Jared Diamond documents the experience of cultures that failed to recognize that their forms of intelligence were unable to understand the emergent, relational, and co-dependent nature of the ecological systems they were dependent upon. The vocabularies that support the West's way of understanding progress as a process of emancipation from traditions, and for the market liberals/libertarians as leading to ever more profits, are also lacking what is distinctive about the different cultural expressions of ecological intelligence.

Certain changes need to be considered in becoming part of the effort to slow the rate of environmental degradation and in recognizing how the globalization of the West's modernizing and progressive agenda continues its messianic tradition of economic and technological colonization. Before turning to discuss this, it needs to be emphasized that what separates today's social justice educators from the market liberal/libertarian promoters of consumerism, and a lifestyle increasingly dependent upon the digital technologies that have already put us on the slippery slope to a techno-fascist future, is that those in the market liberal/libertarian tradition of thinking are focused on increasing profits, exploiting workers, and on overturning civil liberty traditions essential to democracies. The social justice educators, while taking for granted the same emancipatory/progress-oriented vocabulary, are critical of how this vocabulary is being used by market liberals/libertarians to justify the exploitation of others.

Toward a Vocabulary that Reframes How to Think About Educational Reforms that Slow the Rate of Environmental Degradation and the Emergent Techno-Fascist Future

As educational reformers begin to recognize that the warnings of environmental scientists must now be taken seriously, they are likely to be caught in the same conceptual double bind as the 60 or so percent of the public that now acknowledges being concerned about what the future holds for them—but are unable to consider the nature of the lifestyle changes that must be undertaken. The double bind is rooted in being educated to think of themselves as autonomous individuals, in an environment they have traditionally exploited, and in a world of unending progress. That is, awareness that the ecological crisis is also a cultural crisis, where people cannot recognize alternatives to the misconceptions that are at the core of the Enlightenment progress-oriented paradigm, confronts curriculum theorists with the same double bind.

The way out of this double bind, where the long-held taken-for-granted traditions of progressive thinking deepen the crisis, is to begin to think within an ecologically informed paradigm that takes account of how all forms of life are emergent, relational, co-dependent, and participants in different ecologies of information (semiotic) exchange systems. One of the characteristics of understanding that we do not live in a world of autonomous entities and print-based abstractions such as the idea of progress, as the current paradigm holds, is that ecologies of cell behavior, of patterns of insect communication, personal identities, oral and print-based communication, ideologies that justify exploiting workers, and so on, have a history. All ecologies involve observable patterns and relational networks of communication, and it is these observable connecting patterns that provide evidence of whether the ecology is headed in a sustainable direction, or is in a self-destructive mode. Relying upon an ecological paradigm as the source of knowledge means giving close attention to the emergent nature of lived cultural and environmental patterns rather than relying upon the printed word which generally overlooks the taken-for-granted interpretations of the writer and later the reader—as well as the interpretation of classroom teachers and professors who are often unaware of their own taken-for-granted assumptions.

The vocabulary that supports the exercise of ecologically informed intelligence and intergenerational knowledge includes the following: **ecological sustainability, ecological intelligence, intergenerational knowledge and skills, traditions of social justice, indigenous knowledge and skills, wisdom, critical inquiry leading to what should be changed and conserved, double-bind thinking, cultural/bio conservatism, non-monetized relationships and practices, face-to-face communication, living in an interpreted world, ethnically diverse cultural commons,** and **enclosure of the commons.**

This vocabulary is also relevant for understanding how cultural and natural ecologies differ from the market liberal/libertarian (Enlightenment) paradigm being promoted by computer scientists, engineers, and corporate heads who equate progress with collecting data that can be translated into algorithms that eliminate the need for workers while amassing huge profits. Long-term memory is being replaced by short-term memory; abstract relationships are replacing face-to-face relationships; monetized activities and relationships are leading to digital profiles that are being sold to corporations and governments; intergenerational knowledge and skills essential to viable cultural commons that have smaller ecological footprints are being replaced by the seeming convenience and efficiency of online consuming; and the apparent advantages of the Internet are also undermining privacy and freedom from hackers, cyber attackers, and extremist groups. As governments and corporations enhance their ability to gather data to anticipate possible behaviors, which is now called "predictive policing," and to post millions of photographs of people (without their consent) on the FBI network that goes out to all the nation's police agencies, we are fast losing our civil liberties—which online learning is not likely to bring to the attention of students. How does one explain the willingness of so many people to exchange their privacy for the conveniences of the Internet of things and processes? Perhaps the myth of progress has become a religion that promises salvation from the forces of evil.

What the Vocabulary of an Ecological Paradigm Brings to Awareness of How to Reduce the Human Impact on Natural Systems and to Recovering a Degree of Security from the Increasing Threats of the Digital Revolution

As suggested earlier, an ecological paradigm involves a reversal in how language functions within the Enlightenment paradigm where print-based and thus abstract vocabularies influenced both awareness and interpretations of the ecologies of behaviors and communication encountered in daily life. The abstract vocabulary that represented traditions as backward, women as conceptually limited, autonomous individuals as original thinkers, and now data as objective, led to ignoring the complexity of people's lives that did not fit with the stereotypes of a print-based abstract world. The lack of a vocabulary for understanding that all forms of life are dependent upon robust natural systems, like the inability to recognize how women were being repressed, even led social justice reformers of earlier eras to ignore how the biases encoded in the language framed both awareness and what was ignored. To make the point more directly, academics across the disciplines have taken for granted the core Enlightenment assumptions, yet the majority were well behind other segments of society in recognizing gender bias, and are now in recognizing the threat of the digital revolution to our civil liberties and to the cultural commons that represent community-centered lifestyles that have a smaller ecological footprint.

The role of language within an ecological paradigm, which needs to be introduced to students, can be seen in how giving careful attention to the emergent, relational, and co-dependent patterns in both natural and cultural ecologies leads to reframing the meaning of words in ways that are informed by local contexts and an awareness of the cultural patterns that connect. In contrast to print-based Enlightenment-influenced thinkers who take for granted the autonomous nature of a plant and data, ideas such as freedom, free markets, and property, and other entities that supposedly have an autonomous existence, the Buddhist and deep ecological thinker, Thich Nhat Hanh, explains every aspect of the life world as the emergent, relational, and co-dependent. This is overlooked when the word "flower," for example, is named as an autonomous entity—that is, as a plant.

Following his statement that "nothing can exist by itself alone," he goes on to give an account of what can be observed if our thinking and awareness have not been limited by the misconceptions handed down from the past.

> Looking deeply into a flower, we see that the flower is made of non-flower elements. There is nothing that is not present in the flower. We see sunshine, we see rain, see clouds, we see the earth, and we also see time and space in the flower. A flower, like everything else, is made entirely of non-flower elements. The whole cosmos has come together in order to help the flower manifest herself. The flower is full of everything except one thing: a separate self, a separate identity.
>
> (Thich Nhat Hanh, 2002, pp. 47–48)

All aspects of both natural and cultural ecologies can be described in the same way.

The use of a metaphorical language whose meaning was framed by the analogs settled on in the past and passed forward as the taken-for-granted way of thinking, such as the ideas of progress and emancipation from traditions, the objective nature of data, the ecological crisis, and so forth, would all be understood differently if our language were to be informed by an awareness of relationships—and that each of the relationships also has a history that continues to influence the present and even the future. The use of nouns hides the emergent, relational, and co-dependent nature of the ecological world within which we live—which is explained so clearly by Rupert Ross (2006). How the life world is also misrepresented by the use of nouns and pronouns is that the history of words such as "data," "intelligence," "progress," "God," "I," "property," "globalization," and so forth, is ignored.

Curriculum Reforms that Contribute to Exercising Ecological Intelligence

Just as key abstract ideas and traditions, such as encoding knowledge in print and now representing it as data, set the West off on an ecologically unsustainable pathway, there are other key misconceptions that need to be recognized if curriculum reforms are to address the cultural and linguistic roots of the ecological crisis. The following four areas represent the starting points, which need to be expanded as different ethnic groups begin to assess which of their traditions will contribute to slowing the rate of environmental degradation and which have been imposed on them by the colonizing efforts of the West.

(A)   *Promoting Relational Thinking*

As will be explained later, there are many language processes that undermine relational thinking and thus ecological intelligence. Classroom teachers can partly overcome the current tendency of reinforcing what can be called the West's ontology of isolated entities, things, and events that are given a sense of conceptual coherence by relying upon the taken-for-granted interpretive framework acquired from significant others who were, in turn, socialized to think in the metaphorical and largely noun-based vocabulary inherited from the past. The ontology (what is represented as reality) of an emergent, relational, and co-dependent world can be reinforced by asking students in the early grades to consider the experiential differences between the printed and spoken word, between sharing a meal with their family and friends and eating industrially processed food, between face-to-face conversations with friends and communicating through an iPad or cell phone, between caring for a garden and purchasing industrially processed food, between learning a skill from a mentor in one of the arts and being a consumer of someone else's talents. Learning to give explicit attention to the emergent, relational, and co-dependent patterns that connect in the above examples reinforces the exercise of ecological intelligence. As learning to think relationally is undermined by other taken-for-granted patterns in the culture, such as being told by the teacher to figure things out for yourself or construct your own idea (which reinforces the myth of being an autonomous individual thinking about the external world), it will be necessary for teachers to pose questions that prompt students to consider aspects of their taken-for-granted experience that previously were ignored.

Promoting relational thinking in the later grades can involve a wider range of examples, from having students consider the environmental impact of hundreds of thousands of people driving to see their favorite football team, to considering the relationship between the history of scientific achievements and the growing number of prominent scientists now engaging in scientism, to examining who benefits from reducing people's experience to data and from the growth of surveillance technologies, to how the use of printed maps and treaties were part of the colonizing technologies, to how the industrial system of production, now driven by digital technologies that accelerate the process of automation and cultural change, contributes to high levels of unemployment that, in turn, fosters political extremism, to how online consumerism leads to the use of more delivery trucks that release more carbon dioxide into the atmosphere, to where we got the idea of progress and that technologies are both an inherently progressive force and, at the same time, culturally neutral.

Relational thinking needs to be promoted by using examples from the different cultures represented in the classroom, and even from gang cultures. For example, students can be asked to examine the relationships between racial differences, levels of unemployment, prison populations, and the privatization of the prison systems. Other relationships that need to be explored include how the increased reliance upon the Internet affects intergenerational face-to-face communication and thus the forms of knowledge and skills that enable people to be less reliant on the money economy—which leads to exploring alternative economies and cultural differences in how wealth is understood, to considering how the previously invisible minority cultural groups are now represented in television sitcoms and in advertising as driven by the same pursuit of fun, silliness, and consumerism as the members of the dominant white culture. Making relational thinking a part of the curriculum can focus on the mundane, on what students want to explore, and even the deeply important cultural issues and relationships that may not have occurred to students as affecting their future well-being. As prior socialization to the autonomous world of things and stand-alone entities reinforced by the noun-dominated nature of the English language still dominates most students' taken-for-granted world, it is necessary to continually remind students that thinking relationally is part of learning to exercise ecological intelligence—of learning to recognize the patterns that connect within the emergent, relational, and co-dependent worlds of the cultural and natural ecologies that their futures depend upon.

(B)   Demystifying Language Issues

Making explicit the misconceptions about the use of language also involves learning to think relationally—which needs to be explained to students. But here the focus should be on three areas of misconceptions that are particularly relevant to understanding how they affect power relationships and why the ontology of abstract thinking continues to be an ecologically destructive and colonizing force. These include the failure to understand the metaphorical nature of our largely taken-for-granted vocabularies, how print and now data (for all their important uses) reinforce abstract thinking that undermines the exercise of ecological intelligence, and how face-to-face intergenerational communication is essential to revitalizing the cultural commons (which will be explained later).

Understanding the metaphorical nature of language is especially important to becoming aware of how our taken-for-granted vocabularies carry forward the misconceptions and silences that are at the root of so many of the ecological and social justice problems we now face. The question for many Americans, including classroom teachers and university professors, is: How have the misconceptions that fail to represent all living systems as relational, emergent, and as networks of information exchanges led to another misconception that we chose the words that express our own ideas and that represent the nature of the external world of facts and objective knowledge. Why should we take seriously what seems like an absurd and difficult to understand explanation of the metaphorical nature of our taken-for-granted vocabularies? As I have written extensively on the nature and importance of understanding the metaphorical nature of our vocabularies, including why Friedrich Nietzsche got it right and George Lakoff and Mark Johnson got it wrong, I will provide here only a summary of key points (Bowers, 2011, pp. 140–155; 2013, 41–62).

The feminist movement, the shift in thinking of the environment as an endlessly exploitable resource, and the growing awareness that the digital revolution is bringing about cultural changes that do not fit the old understanding of progress provide examples of three metaphors that encode the thinking of earlier eras. That is, the meaning of the words "woman", "environment," and "progress" were framed by the analogs (what something is like) settled upon by influential thinkers in the past who were carrying forward the misconceptions and silences of their era. For example, the *Book of Genesis* provided the analogs for understanding the subordinate role of women. The analogs that framed the meaning of the word "progress," turning it into a universal myth, were derived from the advances in print, from the early stages of modern science, and more efficient and profitable technologies that became the analogs for the mechanistic interpretative framework (root metaphor) for understanding even organic processes—including the human brain.

The taken-for-granted meaning of most of our vocabulary ranging from "civilization," "tradition," "primitive," "individualism," "data," "work," "poverty," "mankind," "God," "science, "technology," and so forth, were framed by the analogs settled upon in the past—and intergenerationally reproduced as new members of the language community relied upon the meaning of words framed by the analogs settled upon in the past. That the analogs that frame the meanings of the vocabulary can be changed is usually not explained, even though feminists and environmentalists continue to introduce different analogs that highlight what is problematic about the old analogs.

Relational thinking can be made explicit as part of helping students examine the nature of the analogs that frame the meaning of words they otherwise take for granted. The political nature of the taken-for-granted metaphorically encoded vocabulary can be seen in how different words privilege certain groups over others, which is now being recognized in terms of gender, ethnic, and racial differences. Relational thinking is also involved in examining how the taken-for-granted vocabularies of market liberals/libertarians prevents them from recognizing that there is an ecological crisis and that it is leading to a catastrophic endgame. Exploring how the use of nouns both serves to hide the metaphorical nature of most words but also marginalizes awareness that life-forming, sustaining, and destroying processes are emergent, relational, and co-dependent, will lead to other important insights.

(C)   How Print and Data Undermine Awareness of the Emergent, Relational, and Co-Dependent World in Which We Live

The taken-for-granted view of print-based cultural storage and communication, which is now being replicated in how the authority of data is being understood, has been focused on the positive contributions of these two Janus technologies. How they both reproduce the Enlightenment view of individual intelligence, a human-centered world, and equate change with progress is less recognized. And how they also reinforce the tradition of abstract thinking that undermines an awareness that we live in emergent, relational, and co-dependent cultural and natural ecologies. As the benefits are well understood, the focus here will be on what has generally been ignored. Again, as I have written extensively on how print and now data undermine the exercise of ecological intelligence that will enable people to recognize how to live less consumer and environmentally exploitive lifestyles, I will only list here what has generally been overlooked about the limitations of print and data (2012, pp. 71–106; 2016). The following points not only need to be discussed by students, but also subjected to a deep ethnographically informed examination of what aspects of their own experience cannot be fully represented in print and by data.

In order to understand the overlooked limitations of print and data it needs to be kept in mind that impermanence, rather than fixed and autonomous entities, characterizes all life-forming and sustaining processes. Print and data provide an abstract understanding, which includes the following: a) both provide an account that provides only a surface knowledge that lacks depth in representing local ecological contexts; b) printed and data-based accounts provide only a snapshot of the flow of experience (which can be tested by obtaining a printed account of a crashing wave or an ongoing conversation); c) what is committed to print, even when used by a gifted writer, too often takes on a life of its own and becomes reified as a universal, which can be seen in the abstract theories of Western philosophers and social theorists; d) the abstract thinking reinforced by print and data-based accounts is inherently ethnocentric as it ignores the emergent, relational, and semiotically complex networks of communication taken into account in oral cultures (that is, face-to-face communication often involves historical memory, awareness of what is being communicated by the Other, critical thought, awareness of traditions, and even empathy); e) what is committed to print and represented as data encodes the taken-for-granted assumptions, cultural interpretative frameworks, and silences acquired earlier in the writer's and data collector's process of primary socialization to thinking in the language handed down from the past; f) because of the limitations accompanying the use of print and data, and the cultural tradition of thinking of language as part of a conduit, that is, a sender/receiver process of communication, both the printed word and data are too often assumed to represent objective facts, information, and data; g) the lack of understanding that the taken-for-granted meaning of most words were framed by the analogs settled upon in earlier times, along with the cultural convention of writing as a third person observer, leads to the widespread failure to recognize that what is written is always an interpretation, and the reader's relationship to what is written or represented as data is also an interpretation based on the taken-for-granted thinking of earlier generations; h) the abstract thinking reinforced by print and data leads to unequal power relationships, especially given other cultural assumptions such as when print is assumed to be evidence of a more rational and advanced civilization than oral cultures. This can be seen in how the use of maps, printed treaties, and the use of Western metaphors established ownership of the lands of indigenous cultures (Newcomb, 2008).

(D)   *Toward Ecologically Sustainable and Community-Centered Lifestyles*

Ecologically sustainable community-centered lifestyles also represent zones of safety from the predatory practices of the hackers, scammers, and surveillance systems that now exist in communities throughout the world. They are called the cultural commons that enable people to be less dependent upon a money economy and the industrial systems of production and consumerism based on the myth of unending progress. The intergenerational knowledge and skills (traditions) passed forward primarily in face-to-face and in mentoring relationships cover the entire range of cultural activities: the growing and preparation of food; ceremonies; the arts including music, dance, and poetry; traditions of social justice; mentoring in how to exercise ecological intelligence; vocabularies; games; craft knowledge of how to work with wood, clay, stone, and metal; and how to read what is being communicated between the natural and cultural ecological systems.

Learning through careful observation of how talents and skills are nourished within the community, how to acquire the skills connected with different cultural commons activities, and how cultural commons activities involve patterns of mutual support, should be a central focus of curriculum reform. The curriculum should encourage students to consider why cultural commons activities are less environmentally destructive than consumerism, and how they lead to being less dependent upon a money economy that will become increasingly restricted as digital technologies and the combination of market liberalism and Enlightenment ideology replace more workers with robots and algorithms. There should also be an in-depth discussion of the relational and co-dependent nature of how the cultural commons conserves traditions of local decision making, while enabling people to be less vulnerable to how digital technologies put at risk their need for security—including their moral narratives central to their sense of cultural identity. Wealth in the cultural commons is understood as the talents and skills that contribute to the well-being of others, and unlike the wealth that is measured in money, it is largely immune from being hacked.

The curriculum should also introduce students to how the ideology of market liberalism/libertarianism continues to undermine what remains of the cultural commons of different cultures, as well as how according high status to print and digital-based knowledge serves to undermine the cultural commons. This is where the earlier discussion of how the vocabulary that supports the myth of progress needs to be reintroduced as part of the discussion of why so many people are unable to recognize that the traditions of today's cultural commons represent alternatives to the industrial/market driven/consumer-dependent culture that is leading the world to the endgame of collapsing natural systems. This discussion should also introduce students to the many groups and movements that go by different names, such as the Transition Communities in the southwest of England, the Business Alliance for Local Living Economies, and the global spread of the Localism Movement for community action as described in the following: www.localfutures.org/wp-content/uploads/climate-action-paper.pdf.

In addition to making different aspects of the cultural commons the focus of ethnographic studies of the community, and exploring issues related to the health of cultural commons activists as well as their satisfaction of living lives characterized by voluntary simplicity, students need to experience the difference between engaging in cultural commons activity and a similar activity that involves a consumer relationship. What are the basic differences in terms of discovering a personal talent and developing the skills that reduce dependency upon consumerism? Many students are already involved in the creative arts, in helping others in the community, and in social justice activism—including environmental restoration projects. Their insights about the experiential differences between learning a skill and participating with others in largely non-monetized activities will help bring out what is ecologically sustainable about the cultural commons.

References

Bowers. C. 2011. *University Reform in an Era of Global Warming*. Eugene, OR: Eco-Justice Press.

_____. 2012. *The Way Forward: Educational Reforms that Focus on the Cultural Commons and the Linguistic Roots of the Ecological/Cultural Crises*. Eugene, OR: Eco-Justice Press.

_____. 2013. *In the Grip of the Past: Educational Reforms that Address What Should be Changed and What Should be Conserved*. Eugene, OR: Eco-Justice Press.

_____. 2016. *Digital Detachment: How Computer Culture Undermines Democracy*. New York: Routledge.

Dewey, J. 1916. *Democracy and Education*. New York: Macmillan.

_____. 1920. *Reconstruction in Philosophy*. New York: H. Holt.

Diamond, J. 2005. *Collapse: How Societies Choose to Fail or Succeed*. New York: Viking.

Freire, P. 1973. *Education for Critical Consciousness*. New York: The Seabury Press.

Havelock, E. 1982. *The Literate Revolution in Greece and its Cultural Consequences*. Princeton, NJ: Princeton University Press.

Lakoff, G., and M. Johnson. 1999. *Philosophy in the Flesh: The Embodied Mind and Its Challenge to Western Thought*. New York: Basic Books.

Newcomb, S. 2008. *Pagans in the Promised Land: Decoding the Doctrine of Discovery*. Colorado: Fulcrum Press.

Nietzsche, F. 1967. *The Will to Power.* New York: Vintage.

Ross, R. 2006. *Returning to the Teachings: Exploring Aboriginal Justice.* Toronto: Penguin Canada.

Sagan, C. 1997. *The Demon-Haunted World: Science as a Candle in the Dark*: New York: Random House.

Thich Nhat Hanh. 2002. *No Death, No Fear.* New York: Riverhead Books.

Part III
Clarifying the Difference between Individual and Ecological Intelligence

10 The Challenge Facing Educational Reformers

Making the Transition from Individual to Ecological Intelligence

The problem today is that most of us have been educated in western-style educational institutions. This has led to being socialized to think and communicate in the metaphorical language framed by analogs settled upon by earlier western thinkers who were unaware of environmental limits. The combination of hubris and deeply held prejudices passed forward by earlier generations toward indigenous cultures that had already developed ecological intelligence that enabled many of them to live within the limits and possibilities of their bioregion led western thinkers to take a different path into the future. As we can now recognize, this path has led to environmentally destructive technologies and the hyper-consumer dependent lifestyle that is now being globalized. Whether we have the time to develop life-sustaining ecological intelligence will depend upon the length of time we have before the rate and scale of environmental changes embroil all Americans in the struggle for survival that will go beyond the current efforts to maintain a debt-dependent standard of living. It will also depend upon whether public school teachers and university professors have the will to recognize how the past continues the linguistic colonization of the present.

Unfortunately, even if our educational institutions are able to socialize the next generations in how to exercise ecological intelligence in their daily lives that goes beyond the pursuit of self-interest, political power will remain in the hands of the older generations who were socialized to the industrial/capitalist mode of consciousness. Even in the face of mounting evidence that the environmental crisis is not a scare tactic of liberals, a large and powerful segment of the American population still places profits and the exploitation of the environment and other people (especially low-wage workers from other countries) above all other considerations. If we are to take Albert Einstein's warning seriously, namely that we cannot rely upon the same mind-set that created the problem to fix it, we need to begin thinking of how to exercise a life enhancing ecological intelligence and thus to move to a post-industrial form of consciousness. This will be an especially difficult challenge for classroom teachers and university professors who have been socialized to think in the metaphorical language that earlier thinkers succeeded in establishing as the basis of modern thought.

As the word "ecology," especially among scientists, has become as widely used as the word "sustainability," it is necessary to identify how it reframes the meaning of "intelligence." Daniel Goleman's book, *Ecological Intelligence* (2009), will likely popularize the phrase among the general population. But his book is unlikely to lead to the realization that ecological intelligence requires a radical shift in thinking that goes beyond being a more informed shopper. Goleman starts off by recognizing that ecologies are complex interacting natural systems that sustain life—including how the future of humans is dependent upon understanding how their behavior impacts the self-renewing capacity of these systems. Unfortunately, he goes on to promote a narrow and basically misleading understanding of ecological intelligence. By reducing it to basing consumer decisions on knowing the life-cycle assessment of various products (that is, the history of the production process and use of toxic materials), any hint that ecological intelligence will require rethinking the widely held view of intelligence as an individual attribute is overwhelmed by his association of ecological intelligence with being a more informed consumer. One of the unfortunate consequences of his book is that many people will begin to think of ecological intelligence as a matter of obtaining information from websites, such as the Berkeley-based GoodGuide, that alert consumers about the ecological impacts of different products. This criticism is not to marginalize the importance of learning about the toxic consequences of different products on various natural systems, but it needs to alert us to the need for a deeper understanding of the nature of ecological intelligence—as well as the modern ways of thinking that undermine it.

The three interconnected areas we need to rethink if educational reforms are to contribute to making the transition to an ecological form of intelligence include the following. First, we need to become aware of the many ways the limited exercise of ecological intelligence is undermined by reinforcing students to think of themselves are autonomous agents. Second, there needs to be wider understanding on the part of educators of how language carries forward the misconceptions and values of earlier thinkers who were unaware of environmental limits. Third, public schools and universities need to introduce students to the ecological importance of revitalizing the cultural and environmental commons, as well as to help them understand how the cultural commons are being undermined by market and technological forces. At the university level, the focus needs to be on the various cultural forces that are transforming what remains of the world's diversity of cultural commons into new markets. These forces include the destructive influence of western philosophers and social theorists who ignored environmental limits, as well as other cultural ways of knowing. Educational reforms also need to focus on how ideologies and religions justify cultural colonization, technologies—such as computers—that marginalize awareness of the importance of contexts and intergenerational knowledge of how to live less consumer-dependent lives, various status systems, and the educational sources of cultural amnesia.

In the interest of brevity, the following summary will focus on key ideas in three areas that must be addressed in thinking about educational reforms that foster ecological intelligence. Thus, the focus will be on the ecologically problematic cultural assumptions and linguistic patterns that are taken for granted by most classroom teachers and university professors and not on the daunting challenge of how to get them to rethink the assumptions their academic careers are based upon.

Fostering Ecological Intelligence

The ancient Greek word *oikos* referred to a wide range of cultural practices in the household and the community. It was only later that Ernst Haeckel (1834–1919) transformed it into the neologism "oecologie" that eventually became "ecology." He then identified it as the study of natural systems. We need to recover the ancient Greek understanding of learning the cultural patterns of moral reciprocity essential to community while also retaining the more contemporary understanding of the behavior of natural systems as ecologies. Both cultural and natural ecologies involve interdependent systems, where no organism or action exists on its own. Gregory Bateson refers to the changes circulating within different ecosystems, and within and between cultural and natural systems, as the "difference which makes a difference." (1972, p. 315) These differences, or actions upon an action, can also be understood as the patterns that connect, which in turn lead to changes in other parts of the cultural and natural ecologies. In short, ecological intelligence takes account of relationships and contexts, as well as the impact of ideas and behaviors on other participants in the cultural and natural systems. Rachel Carson's recognition of the connections between the use of DDT and the decline in the local bird population is an example of recognizing the interactive patterns that are not always life enhancing. Many of her critics took for granted that, like other scientific discoveries, DDT was yet another expression of progress—which led them to ignore its impact on natural systems. The myth of progress, especially scientific-based progress, reinforced the taken for granted pattern of thinking that, in turn, led to ignoring the difference (introduction of a pesticide) that makes a difference (the dying off of birds).

Ecological intelligence is what many indigenous cultures rely upon in order to adapt their cultural practices to the cycles of renewal in their bioregions. For example, the Quechua of the Peruvian Andes express ecological intelligence in their ability to observe what the changes in their environment are communicating about when and where they should plant their fields. Their ceremonies both re-enact the patterns of human/nature interdependence and give thanks for how nature nurtures them. Ecologically oriented scientists are now exercising a limited form of ecological intelligence as they study the energy flows and cycles of renewal. Unfortunately, they continue to ignore the deep cultural patterns of thinking, including how these patterns are encoded in the language they take for granted. Social scientists also rely upon a limited form of ecological intelligence when they study the patterns that connect, such as how the patterns of discrimination and class differences impact the lives of people. What they overlook is how the natural ecologies are undermined by the cultural assumptions that underlie their individually centered way of thinking of social justice.

Ecological intelligence takes into account the interacting patterns, ranging from how behaviors ripple through the field of social relationships in ways that introduce changes that are ignored by non-ecological thinking, to how an individual's actions introduce changes in the energy flows and alters the patterns of interdependence within natural systems. When we pay attention to contexts, interactions, and the consequences that follow from these actions, we are also exercising ecological intelligence. Ecological intelligence is not something we have to create anew, as it goes back to the form of intelligence exercised in hunter-gatherer cultures. They had their mythopoetic narratives, but their survival depended upon careful observation of the cycles and patterns in the environment—as well as the intergenerational knowledge they continually tested, refined, and encoded in their technologies and language systems.

Unfortunately, western philosophers from Plato to the present have largely denigrated this form of intelligence by representing rational, abstract, and thus decontextualized thinking as having a higher status. Over the centuries, ecological intelligence has been further undermined as the idea of the autonomous individual became accepted as the basis of our political and social justice system, and now as the source of ideas and values. The introduction of perspective by artists in the 15th century helped to strengthen the cultural myth that privileged the individual as a separate onlooker on an external world, just as René Descartes further strengthened the myth of intelligence as separate from the cultural and natural ecologies that individuals interact with in ways that are too often ignored. Today, the myth of the autonomous individual is being reinforced by educators who urge students to construct their own ideas, and who promote computer-mediated learning on the grounds that it enables students to decide what they want to learn and value. Cell phones, as well as many other cultural forces, further undermine awareness of contexts, relationships, interdependencies, and the consequences of human behaviors that ripple through both cultural and environmental ecologies. Such taken for granted linguistic conventions as using the phrases "I think," "I want," and "What do you think?" continually reinforce the myth of not being part of the interdependent cultural and natural ecosystems, but rather being a separate observer, thinker, and actor.

What are the implications for educational reformers? The first would be to become more aware of how the taken for granted cultural assumptions influence whether intelligence is interpreted as the attribute of an autonomous individual. Special attention needs to be given to how the student may represent her/himself as being an autonomous observer, and as a source of originality and intentionality. As noted above, this assumption, as well as many of today's other taken for granted cultural assumptions, gave conceptual direction and moral legitimacy to the industrial/consumer culture that is now entering its digital phase of globalization. Other assumptions include the idea that change leads to a linear form of progress, that this is a human-centered universe, that mechanism provides the best explanatory framework for understanding organic processes, that language is a conduit in a sender/receiver process of communication, that traditions limit the individual's freedom and self-discovery, that (still for some) patriarchy was part of the original creative process, and that free markets have the same universal standing as the law of gravity.

A second suggestion would be for participants in a learning situation to reinforce each other in giving greater attention to the cultural and environmental patterns that connect, to the consequences that follow from different behaviors, and to whether these consequences have an empowering or detrimental effect on others—in both the cultural and natural systems. The subjectively centered self is such a prominent tradition in mainstream western culture, even among artists and people searching for a deeper sense of meaning and purpose, that it needs to be discussed and, if possible, reframed in ways that take account of how an action affects the actions of others, including the natural systems, in ways that influence their development. A key to making the transition to ecological intelligence is recognizing that there are no isolated events, facts, or actions. Everything, as Bateson points out, is part of a larger system of information exchanges. One of the more difficult sources of resistance to obtaining this awareness is the way in which print, both in books and in computer-mediated communication and thinking, marginalizes the importance of contexts, tacit understandings, and awareness of the history of the larger network of relationships. Even when what is represented by the printed word is situated in terms of its history, the print-based historical account is also an abstract construction that is unable to accurately represent the culturally mediated embodied experiences of participating in the cultural and natural ecology of an earlier time.

How Language Influences Our Thoughts and Values in Accordance with the taken for Granted Assumptions of the Culture

Just as the cultural assumptions have led to thinking that individuals are basically autonomous beings (or have the potential to become autonomous), we have a tradition of thinking of the other participants in these complex cultural and natural ecologies as being self-contained entities, such as a weed, a crime, a behavior, a value, an idea, and even the printed word. The spoken word, on the other hand, makes it easier to recognize the different dimensions of the cultural ecology in which it occurs. Context, memory, reciprocal actions, tacit understandings, and immediate consequences are accessed through all the senses and affect understandings and actions. Given the privileged status that the printed word has in public schools and universities, it is necessary to emphasize the importance of helping students recognize that words are not autonomous entities into which teachers/professors, authors, and computer software writers put their meanings and then convey them to others.

Our educational institutions leave most graduates with the idea that language is a neutral conduit that enables ideas, objective data, and information to be passed to others. That is, the majority of students graduate without understanding that most words are metaphors that carry forward the meanings framed by an earlier choice of analogs. Many of these analogs were chosen by men who were unaware of environmental limits, and who took for granted many of the cultural assumptions of their era. Recognizing that words have a history has important implications that are seldom considered. That is, they are part of a complex linguistic ecology that can be traced back to a culture's earlier narratives and evocative experiences. Thus, the use of such words and phrases as tradition, technology, property, data, intelligence, progress, critical thinking, and so forth, carry forward the way of thinking of earlier times—as well as the silences and prejudices that were taken for granted within the culture.

Overcoming this general lack of historical perspective suggests one of the ways in which educational reformers can foster ecological intelligence. Students need to be encouraged to examine the history of key words in the modern vocabulary that are contributing to undermining the intergenerational knowledge of the community, to the colonization of other cultures, and that lead to behaviors that further degrade the environment. For example, they need to consider the cultural context that influenced John Locke's analogies for understanding the right of individuals to own property, and the early cultural basis for thinking of technology as a neutral tool, as well as the basis for thinking of traditions as obstructing progress and rational thought.

Ecological intelligence involves escaping from the linguistic colonization of the present by the past. An especially critical example of when ecological intelligence needs to be encouraged is when professors, political elites, and demagogues in the United States refer to the economic policies of Milton Friedman, Ronald Reagan, and George W. Bush as conservative. These promoters of capitalism and the globalization of free markets should be more accurately described as market liberals. To reiterate a key point: words have a history, and the word conservative, when used as a category of political theory in the West, can be traced back to Edmund Burke who warned about the danger of basing changes on abstract (and supposedly universal) ideas, to Michael Oakeshott who explained how the rationalization of the workplace de-skills the worker, to the authors of *The Federalist Papers* who justified the separation of powers, and to contemporary environmental thinkers such as Aldo Leopold, Wendell Berry, and Vandana Shiva.

The word conservative carries forward many problematic interpretations of what should be conserved, such as states' rights and prejudicial traditions.

The important point, however, is that the genealogy of political metaphors such as conservatism, liberalism, libertarianism, socialism, and Marxism, as well as the root metaphors that frame their respective agendas and silences, need to be examined in terms of their hidden forms of cultural colonization. Given the current threat in the United States to our civil liberties, including such long-standing traditions as habeas corpus, democratic practices, and to the self-renewing capacity of natural systems, it is important to think ecologically about how to rectify the use of our political vocabulary that may otherwise lead people to equate the political slippery slope leading to the further enclosure of the local cultural commons with modern progress and development. What Naomi Klein documents in her recent book, *The Shock Doctrine: The Rise of Disaster Capitalism* (2007), is a powerful example of how modern political metaphors hide the process of economic and cultural colonization. Ecological intelligence avoids accepting the authority of abstract words and theories by focusing on how the consequences of policies affect the prospects of the other participants in the larger cultural ecology—as well as on the fate of the natural systems.

How Fostering Ecological Intelligence Leads to Revitalizing the Local Cultural Commons

To reiterate, the way we unknowingly accept basing relationships and values on the meaning of words that were framed by analogs selected hundreds of years ago becomes especially critical to whether we can make the transition to a post-industrial form of consciousness and community. Substituting the phrase cultural and environmental commons for what most people associate with the word community will help in making this transition. Even in its most positive use, the word community is too limited to convey the complexity of the cultural and natural ecologies upon which we depend. Stripped down to the simplest explanation, the cultural commons represents the intergenerational knowledge, skills, and mentoring relationships that enable community members to be more self-reliant in the areas of food, healing, creative arts, craft skills, narratives, ceremonies, civil liberties, and other aspects of daily life that are less dependent upon consumerism and participation in the traditional money economy. Basically, it encompasses what is shared in common, which may also include traditions of exploitation and prejudice.

The word commons is now being used to refer to the cyber-commons, and its history in understanding the environment as a commons can be traced back to Roman law. The intergenerational knowledge and skills now being widely shared—ranging from how to grow, prepare, and share a meal, and how to discover talents and skills in a wide range of the arts, to the local efforts to make political decisions that protect the local cultural and environmental commons from being integrated into the supposedly free-market economy—have profoundly different consequences than what is experienced in a consumer-dependent lifestyle. Revitalizing the cultural commons enables people to be less dependent upon a money economy that too often exploits the most vulnerable people, as well as the environment that future generations will depend upon.

The intergenerational knowledge and skills that represent alternatives to the industrial mode of production and consumption also have a smaller carbon and toxic ecological footprint. Furthermore, strengthening of the local cultural commons leads to developing the skills and relationships that are the basis of mutual support. In short, these life-sustaining forms of ecological intelligence will vary from culture to culture and from bioregion to bioregion. And like traditions, such as the slow food movement, that are carried forward by mentors, the cultural and environmental commons will continue to exist, along with a more selective dependence upon modern technologies. The challenge for educators, which is being made more daunting by the ideology that justifies greater reliance upon computer-mediated learning, is to help students become aware of the forms of knowledge that take account of the limits and possibilities of the local bioregion, as well as patterns of mutual support that are essential to moving into the post-industrial era that we must enter if we are to avoid total ecological collapse.

The revitalization of local cultural commons across North America, and in Third World cultures that are questioning the western model of development, involve intergenerationally connected relationships and mutual support systems. These relationships, if examined in terms of specific activities and skill development, are not framed in terms of fostering more "individual self-direction," "independence," and "ongoing questioning and revising." These words and phrases are based on the same deep cultural assumptions that led to the kind of individualism required by the industrial/consumer-oriented culture. As these words and phrases have a special standing in the thinking of both market and social justice liberals, it is important to clarify how metaphors that are often associated with progress in achieving fuller individual lives may actually support the forces that lead to a consumer-dependent lifestyle. In *Rebels Against the Future* (1995), Kirkpatrick Sale notes that "it was the task of industrial society to destroy all . . . that 'community' implies—self-sufficiency, mutual aid, morality in the marketplace, stubborn tradition, regulation by custom, organic knowledge instead of mechanistic science . . ." He goes on to identify the connection that is often overlooked by educational reformers who emphasize the importance of individual emancipation: namely, that all the local cultural commons "practices that kept the individual from being a consumer had to be done away with so that the cogs and wheels of an unfettered machine called 'the economy' could operate without interference . . ." (1995, p. 38) In short, the industrial/consumer-oriented culture requires the further enclosure of the cultural commons and an educational system that hides the dynamics of how language, in carrying forward the analogs settled upon by earlier culturally specific thinkers, is part of this colonizing process.

The linguistic tradition of reproducing the conceptual errors of the past (in this case, the analogs settled upon by Enlightenment thinkers) can still be seen in how much of our thinking represent "traditions" as an obstacle to progress and individual self-discovery. However, when we consider the traditions of organic gardening, of craft skills and knowledge, of the creative arts, and of local decision making about how to protect civil liberties and the viability of the environmental commons, we find that the traditions we re-enact and modify in daily life are not always impediments. The contextually grounded nature of ecological intelligence does not require treating progress as in opposition to traditions or the students' discovery of interests and development of talents as being undermined by the forms of intergenerational knowledge and skills that are the basis of mutual support in the community.

If we consider most learning relationships, without succumbing to the meaning of words dictated by the ideology of various expressions of liberal/progressive thinking that have given us a mixed legacy of social justice achievements and the industrial/consumer-dependent culture, we will find that traditions, intergenerational knowledge and skills, awareness of relationships and patterns of mutual support, the use of language that takes account of context and tacit understandings, and moments of dialogue are integral to the students' pursuit of interests, questions, and desire to achieve at a deeper level of accomplishment. We need to continually think against the grain of today's formulaic thinking by keeping in mind that the western theorists who identified the analogs that now frame the meaning of such words as progress, individualism, freedom, emancipation, and so forth, were not aware of ecological limits. Their analogs reflected the advanced thinking of their era. Like the Roman god Janus, their vocabulary enabled us to make important gains in the area of correcting social injustices and in establishing a framework for civil liberties. Now we need to revise this vocabulary in ways that are culturally and ecologically informed. These words can then take on the meanings that reinforce the exercise of ecological intelligence, which requires becoming more ethnographically informed about the differences between the cultural patterns that strengthen traditions of community mutual support and those that adversely impact the viability of natural systems.

References

Bateson, G. 1972. *Steps to an Ecology of Mind.* New York: Ballantine Books.

Goleman, D. 2009. *Ecological Intelligence: How Knowing the Hidden Impacts of What We Buy Can Change Everything.* New York: Broadway Books.

Klein, N. 2007. *The Shock Doctrine: The Rise of Disaster Capitalism.* New York: Metropolitan Books/Henry Holt.

Sale, K. 1995. *Rebels Against the Future: The Luddites and Their War on the Industrial Revolution.* Reading, MA.: Addison-Wesley.

11 Gregory Bateson's Contribution to Understanding Ecological Intelligence

The challenge facing educators is far more complex than merely providing students with the data connected with the scientific findings about changes in the earth's ecosystems. It is also more complex than educating them in how to develop new technologies that are less disruptive to natural systems. As the late Gregory Bateson warns, our survival depends upon a radical transformation of the dominant patterns of thinking in the West. These patterns are widely shared, passed along in everyday conversations, and encoded in the built culture. The institutions that give special legitimacy to these patterns of thinking are the public schools and universities. They have the greatest potential for providing the conceptual space necessary for understanding the historical roots of the misconceptions underlying the myth that if humans rely upon rational thought they can control the changes occurring in natural systems. They also are the sites where students can learn about the nature of ecological intelligence and how the exercise of ecological intelligence leads to correcting the destructive impacts of earlier assumptions and practices on natural systems and human communities. One of Bateson's key insights about the recursive nature of cultural belief systems reminds us that past ways of thinking, both in terms of the conceptual history of the culture as well as the conceptual history of the professor, are likely to be ignored—thus dooming to failure the efforts to correct the conceptual errors of the past.

Before discussing the fundamental differences between the dominant view of individual intelligence (including the cultural assumptions that support it) and the nature of ecological intelligence, a brief sketch of Gregory Bateson's background would be useful. He was born into the family of a prominent British biologist in 1904 and died in 1980. He began as a student of zoology but quickly shifted to the field of anthropology—which led to his fieldwork in New Guinea where he collaborated with Margaret Mead, whom he eventually married and later divorced. According to his own account, his first book, *Naven,* contained his initial insights about the hidden influences on the observer's perceptions and analyses. Following his move to the United States, Bateson began to work in the field of psychotherapy and to participate in the early discussions of cybernetics. Both fields led to important developments in his understanding of the connections between communication processes and what he refers to as the double binds in human and human-and-nature relationships that perpetuate the problems rather than solving them. His last two books, *Steps to an Ecology of Mind* (1972) and *Mind and Nature* (1979), are now recognized as his most important contributions. The former, which is the most widely read, is a collection of essays and printed versions of talks he gave to various audiences. As in nearly all cases where radically new ideas are presented to groups that are encountering them for the first time, the introduction to key ideas and themes tends to be repeated in different sections of the book. The elaboration on certain key ideas, as well as Bateson's arguments with counter points of view require book-length treatment—which can be found in Peter Harries-Jones' excellent book, *A Recursive Vision: Ecological Understanding and Gregory Bateson* (1995).

My purpose here is to introduce several of Bateson's more fundamental ideas and to explain how they lead to rethinking both the idea of individual intelligence and the cultural assumptions that support it. I also explain how his insights are the basis for understanding ecological intelligence, as well as their practical implications for introducing educational reforms that do not rely upon the past misconceptions that are major contributors to putting our culture on an ecologically destructive pathway. While I introduce several ideas from Bateson's other writings, the clearest account of his insights can be found in the sub-section of his chapter on "The Cybernetics of 'Self': A Theory of Alcoholism" (1972, pp. 309–337). The sub-section is titled "The Epistemology of Cybernetics" and is a mere six pages in length. The challenge will be to expand upon his short explanations in a way that enables the reader to recognize how they transform our traditional individually centered understanding of how we acquire knowledge, engage in relationships with others and the environment, and begin to make the transition to an ecological way of thinking. First, I will present a summary of the different ways in which individualism and the supporting cultural patterns are part of the experience of most Westerners. There are, of course, variations in how this sense of individualism is experienced. Differences can be traced to the influence of local cultural traditions, ideologies, religions, and what has been learned from personal experience.

Summary of Assumptions Underlying Being an Autonomous Individual

The personal pronouns "I," "me," and "you," as well as the names we are given set us apart from others, and continually reinforce the sense of being an autonomous individual. This culturally mediated experience is further reinforced by the cultural tradition that emphasizes sight over the other senses as the most accurate way of acquiring knowledge—which reinforces the sense of being separate from the object observed. This leads to the subjective experience of having a unique perspective on events in the external world. The conduit view of language (which I have written about elsewhere) and the idea of objective knowledge promoted by the intellectual class also marginalize the awareness of the cultural and environmental influences that must be ignored if the myth of being an autonomous individual is to be sustained. Taken-for-granted cultural assumptions required to support this myth include the ideas that change is a linear form of progress and that this is a human-centered universe. Also reinforcing the idea of individualism are the Enlightenment assumptions about the power of rational thought. Today, Enlightenment thinkers and their followers have contributed to the widespread cultural amnesia by framing the meaning of the word "tradition" in a way that has reduced it to whatever is associated with maintaining privileges, backwardness, and oppressive practices. The myth of progress, and the increased reliance upon computers which reinforces the idea that the individual is in control of where in cyberspace she/he wants to explore, further adds to a state of awareness that makes traditions appear as irrelevant. In addition to the tradition of civil liberties, which George Lakoff wrongly identifies with progressive thinking, there are overwhelming economic and technological forces that further strengthen the special status that individualism has in western cultures. These include market liberalism, as it equates the expansion of capitalism with the expansion of individual freedom, and libertarianism, as a more extreme ideology that celebrates the "Virtue of Selfishness"—which is the title of one of Ayn Rand's books.

The special status given to print-based technologies, such as books and computers, while having many positive and essential benefits, also reinforces abstract thinking and the individual's ability to exercise critical thought. Critical thinking has led to challenging many sources of oppression and has clearly contributed to important achievements in the area of social justice. However, a more complex understanding of the many uses of critical thinking will reveal that it is also used by special interest groups who are working to overturn government regulation of exploitative practices, and to manipulate public opinion in order to gain support for foreign wars that benefit corporations and the military personnel's need for steady advancement through the ranks. Critical thinking is essential to developing new strategies for manipulating the public's consumer addiction and its willingness to support a bloated military budget. Advertising agencies and various extremist groups also rely on critical thinking to develop their strategies, as do social justice groups—albeit for radically different purposes.

There are cultural traditions that have not been completely marginalized by these various emphases on individualism. The traditions of the natural and cultural commons, while under threat by market forces, now are undergoing renewed support by members of local communities where mutually supportive values and interests are recognized as giving meaning to what is too often the autonomous individual's sense of isolation and lack of meaningful purpose. Other individuals who are working to improve the quality of everyday life by strengthening the community's infrastructure of roads and public services are finding, to quote the title of Robert Putnam and Lewis Feldstein's book, that life is "better together." They are an example of the civic individualism mentioned earlier. These groups, as well as religious groups trying to live by the moral guidelines of the Social Gospel, are motivated by a connected sense of individualism that goes against the grain of market liberal and libertarian thinking. Social justice advocates and environmentalists also have political and moral agendas that differ radically from the larger segment of the population that places self-interest and reliance upon what they assume are their own ideas above all else.

A public school and university education is another powerful force that contributes to the myth of being an autonomous individual—or at least having the capacity to achieve this highest expression of human self-realization. Classroom teachers and university professors have adopted a number of strategies for convincing students that they are accountable for having their own ideas and values. Educators reinforce this message by encouraging students to create their own values, to identify what careers they want to pursue, and to rely upon the wealth of information and data available on the computer for developing their own ideas. At the university level, students are expected to cite the source of ideas that they have not originated. The irony is that this expectation is supported by not informing students about how the language they rely upon to express their "own" ideas is actually metaphorical in nature and thus carries forward the prejudices and silences that were the basis of the taken-for-granted patterns of thinking of earlier generations. Included in the misconceptions reinforced by most faculty are the ideas that the rational process is free of cultural influences, that there is such a thing as objective interpretations, information, and data—and that abstract knowledge is more reliable than what is passed along through face-to-face communication and, generally, oral traditions. The curriculum in most institutions of higher education mirrors that of a supermarket, where individual choice maximizes the appeal of a university education and also has a powerful influence on the students' sense of being autonomous, self-directed agents. The connections between the ways in which individualism is reinforced, and the actual culturally mediated embodied experience of the student, are too complex to be fully addressed here. Nevertheless, this overview is adequate for highlighting why these various expressions of individualism inhibit the development of ecological intelligence. It is also adequate for recognizing why Bateson's ideas lead to a radically different, and indeed more accurate, understanding of human/nature relationships than the explanations provided by the conceptual and moral mainstream western cultures.

Bateson's Insights about the Nature of Ecological Intelligence

Any discussion of Bateson's core ideas is likely to be met with an immediate response of incomprehension and frustration, especially for the reader who has become accustomed to ideas being reduced to little more than sound bites. The following statements, which appear in the six short pages I suggest as providing the best overall introduction, turn out to be two of his most profound insights. They will be more fully explained as we go more deeply into Bateson's other key ideas. Especially important is his statement: "A 'bit' of information is definable as a difference which makes a difference. Such a difference, as it travels and undergoes successive transformation in a circuit, is an elementary idea." (1972, p. 315)

As we shall see, this statement about differences being the basis of the information networks we more conventionally know as an ecology is also critical to understanding the following:

> The total self-corrective unit which processes information, or, as I say, 'thinks' and 'acts' and 'decides,' is a *system* whose boundaries do not at all coincide with the boundaries either of the body or of what is popularly called the 'self' or 'consciousness'; and it is important to notice that there are *multiple* differences between the *thinking system* and the 'self' as popularly conceived.
>
> (1972, p. 319)

The question that might arise in trying to make sense of these two statements is: How did Bateson's education lead him so far astray? And his response would likely be the question: How did the West fail to recognize that it was making a major epistemological error when it emphasized things as separate entities? In the following statement he corrects what he regards as this basic mistake in thinking: ". . . while I can know nothing about any individual thing by itself, I can know something about the *relations between things*." (1987, p. 157)

Part of the answer to why things rather than the relations between things has become a dominant pattern of thinking in the West can be attributed to the privileging of a print-based form of consciousness over that of oral/narrative-based cultures. Plato, according to Eric Havelock, played an important part in this transition, which had the effect, along with many important benefits, of marginalizing the importance of contexts. Without an awareness of contexts the printed word takes on the role of referring to things—which is an abstraction just as the use of the personal pronoun "I" is an abstraction. Individuals, plants, animals, rivers, geological formations, etc., can be represented in terms of their physical characteristics, and even in terms of their behaviors. But this leads to a highly restrictive understanding, one that largely omits the formative relationships and interactive patterns within the larger ecology. Print allows for explanations of causality, but even these represent the writer's interpretation of relationships. The myth of objectivity helps to hide the imposition of the author's interpretation, which may be framed by cultural assumptions of which she/he is seldom aware. These assumptions, in turn, go far back into the past of the language community.

Contemporary examples of this cultural proclivity to think of things as distinct entities rather than the formative influence of their relationships (or what can be referred to as the ecology of which they are a part) can be seen in the way species from other parts of the world have been introduced into different regions of North America—with disastrous consequences for native species. Fields, rivers, forests, and even backyards are now undergoing dramatic changes as native species are being crowded out. This cultural emphasis on separate entities, rather than on formative relationships, can also be seen in how a worker is defined in terms of a salary, a student in terms of a grade, a product in terms of the price put on it—and the way in which an individual's identity can be reduced to a social security number. As Daniel Goleman documents in his book, *Ecological Intelligence* (2009), understanding a product in terms of its life cycle assessment—that is, its production history which includes the use of chemicals, toxins released into the environment, the amount of energy required, the patterns of labor, and the ecological footprint connected with its recycling—represents an alternative to the long tradition of thinking of things in terms of distinct entities. Similarly, in the past the student's grade was assumed to be an expression of her/his intellectual achievement, but recently there is a greater emphasis on considering the formative relations that may be responsible for the student's level of performance. There are many other examples where things are no longer understood in isolation from their surroundings—that is, the larger ecology of relationships in which they participate. Nevertheless, print-based knowledge continues to marginalize contexts, and tacit understanding continues to perpetuate the emphasis on "things," as does our daily practice of relying upon nouns and pronouns rather than verbs.

Equally important are several other ideas that differ radically from the dominant way of thinking in the West—and are critical to understanding Bateson's statement that differences are elementary ideas and sources of information, and that the unit that processes information is much broader and more inclusive than the thinking individual. These include the ideas of recursion, that the map is not the territory, the nature of double bind thinking, that human intelligence and action do not occur as processes separate from the information circulating through the relationships that make up the system, and that in systems that show mental characteristics no part can exert unilateral control over the whole. Each of these ideas needs to be integrated so that they are understood as part of a larger system of ecological intelligence. Ways of thinking that depart from ecological intelligence, as Bateson puts it, lead to an ecology of bad ideas that threatens the system as a whole. Each of these ideas is essential to how Bateson understands the ecology of mind. Introducing each idea separately also makes it possible to identify how thinking of intelligence as one of the distinguishing attributes of the autonomous individual limits her/his awareness of how embodied experiences are nested in larger information networks.

The Many Faces of Recursion—And the One Most Related to Ecological Intelligence

Recursive patterns of thinking exist in a variety of areas—including mathematics, computer science, and in a culture's ways of knowing—or what Bateson refers to as a "recursive epistemology." The focus here will be on understanding what he means by a recursive epistemology and what this phenomenon helps us understand about why educators continue to reinforce the ecologically uninformed patterns of thinking that have their roots in earlier mythopoetic narratives (including the writings of major western philosophers) and powerful evocative experiences—including experiences shaped by technologies mistakenly thought of as neutral "tools."

As in so much of Bateson's writings, ideas are seldom presented in a straightforward manner where the reader obtains what might be considered a final definition and not a further engagement with the ideas of other theorists. The possibility of a workable definition is often sacrificed by Bateson's own qualifications as he rethinks his own insights—and how far they can be generalized. The following list both presents a key feature of Bateson's understanding of recursiveness in the culture/language/thought process, as well as the difficulty of penetrating his conceptual process. In a collection of essays edited by his daughter, Mary Catherine Bateson, and published under the title, *Angels Fear: Toward an Epistemology of the Sacred* (1987), Bateson presents an explanation of recursion as a characteristic of structure, which he extends to the epistemological structure of a cultural way of knowing.

1   'Structure' is an *informational* idea and therefore has its place throughout the whole of biology in the widest sense, from the organization within the virus particle to the phenomena studied by cultural anthropologists.
2   In biology, many regularities are part of—contribute to—their own determination. This *recursiveness* is close to the root of the notion of 'structure.'
3   The information or injunction which I call 'structure' is always *at one remove from its referent*. It is the name, for example, of some characteristic immanent in the referent, or, more precisely, it is the name or description of some relation ideally immanent in the referent.
4   Human languages—especially perhaps those of the West—are peculiar in giving undue emphasis to Separate Things. The emphasis is not upon the relations between but upon the ends of relationship, the relata. This emphasis makes it difficult to keep clearly in mind that the word 'structure' is reserved for discussion of *relations* (especially to be avoided is the plural use 'structures').
5   Insofar as the name is never the thing named and the map is never the territory, '*structure*' is never '*true*.' (1987, p. 161)

Peter Harries-Jones summarized one of the ways in which Bateson's understanding of recursion can be understood: "Recursion as a process of continuous looping [is] a process without observable attributes of structure" (1995, p. 187). The metaphor of "looping" is useful here as it suggests that life-forming and sustaining processes, including life-threatening processes, do not move in the linear direction modern thinkers associate with progress. To stay with Bateson's example, the original conceptual structure that leads to the use of language, and thus to the pattern of thinking that is finely attuned to naming things rather than relations and contexts, is further reinforced when this pattern of thinking is exercised today. To make the point more directly, the conceptual structures (or what I refer elsewhere to as the root metaphors) formed in the past continue to influence the present, and the present loops back to reinforce the conceptual structures formed in the past. For example, the root metaphor of mechanism introduced by Johannes Kepler and other scientists—which enabled them to think of phenomena in terms of measurement, experimentation, and innovation—has become reified and today is the conceptual model for understanding a wide range of processes, including the human brain, the genetic engineering of seeds, and behavior modification.

Other root metaphors inherited from the past continue to frame how people think today—including their use of vocabulary, shared silences, and explanations of causal relationships. These different examples of recursion exist as part of the collective and tacit memory of the culture. Because Bateson often explains recursion by using the concept of a thermostat to make his point about the looping or feedback of information in a self-perpetuating system, some readers have mistakenly assumed that he is promoting a mechanistic way of thinking. This has led to overlooking the more important implications of understanding cultures as recursive systems. Also overlooked is that he is identifying one of the key reasons that we continue to rely upon the same metaphorical language to extricate ourselves from the ecological crisis to which this language contributed.

There are two key insights that Bateson brings to our attention: one being that we take for granted the conceptual structures rooted in the distant past and continually reinforced through the thought patterns of succeeding generations. The other insight is that today's tendency to associate change with linear progress is a cultural construction that loops back and repeats the earlier symbolic structure that came into existence when Enlightenment thinkers interpreted the emergence of modern sciences and a technological form of consciousness, literacy, the idea that rational thought should replace traditions, and the idea that humans were not only given the power to name the participants in the natural world but also to exploit them for their own purposes.

Bateson does not ignore other aspects of the West's recursive cultural epistemology. His criticisms include all the characteristics of modern consciousness that can be traced back to the earliest mythopoetic narratives that focused attention on the abstract religious debates about what follows the death of the individual rather than on learning from the behavior of the natural systems upon which people depend. Other recursive patterns underlying modern consciousness include powerful evocative experiences, such as organizing daily life in accordance with the rhythms of a mechanical clock and the many ways of representing and justifying a linear view of progress. Many of Bateson's most direct and explicit criticisms are directed at the Cartesian view of the individual that represents thinking and awareness as separate from the world of interacting relationships.

Among the recursive patterns classroom teachers reinforce is the idea that individuals have the power to originate their own ideas. This is reinforced when the teacher asks: What do *you* think? What do *you* see (where it is assumed that the student has a unique vantage point on the external world)? What do *you* want to happen? Who do *you* want to become as an adult? And so forth. The emphasis on nurturing the student's creativity, experimentation, and even achieving the fullest expression of freedom by rewriting the ends of traditional stories in ways more in line with what the student values and wants to happen, are all common examples of what is reinforced in classrooms— especially in the earliest grades. The recent emphasis on students constructing their own ideas, relying increasingly upon computers as a way of accessing abstract information, explanations, and simulations, and now the addiction of students to communicating through cell phones and text messaging, all reinforce the view of the individual who, as Descartes announced, possesses the power to exercise rational thought that is free of the influence of traditions. This is, of course, an illusion promoted by progressive-oriented cultural forces that too often have given legitimacy to replacing the non-monetized traditions with reliance upon consuming goods and services. However, unlike the Santa Claus illusion, most people never wake up to the reality of how many daily traditions they rely upon.

The irony is that while some classroom teachers are encouraging students to be more conscious about relying upon local sources of food and engaging in recycling, they are still reinforcing the abstraction that represents the individual as autonomous—or at least has the potential to become so. Even the more ecologically informed approaches taken in environmental education classes fail to challenge the Cartesian misconception that represents the individual as an independent observer of phenomena occurring in the local streams and other environmental sites. To give this criticism greater credibility, one has only to look at the failure of teachers of environmental education to introduce students to the idea that the words they use have a history, and that these words—such as progress, technology, community, science, and so forth—involve the recursive process of repeating the same silences and misconceptions of earlier thinkers who also took for granted the conceptual structures of their culture and era.

With the major exceptions being in the sciences that have taken an ecological turn, most university faculty also reinforce the idea of the student as an autonomous entity who is responsible for making explicit the distinction between her/his own ideas and those derived from outside authorities. Also reinforced is the use of personal pronouns that reinforce the myth of individual autonomy, as well as the emphasis on objective knowledge that hides the influence of the languaging processes that reproduce the recursive interpretive frameworks currently taken for granted or expressed in the new vocabulary of critical thinkers. The result is that most students graduate from colleges and universities thinking of language as a conduit in a sender/receiver process of communication and that what is shared in this sender/receiver process of communication are their own thoughts and values. Few are aware that their use of language and their thoughts repeat the earlier deep patterns of thinking formed before there was an awareness that many non-western cultures had prioritized the importance of understanding relations within natural systems and had developed an ecological form of intelligence, an awareness that there are environmental limits, and an awareness that the printed word is profoundly different from the living nature of the spoken word.

Just as the public school teacher who continually recycles the myth of linear progress as a way of justifying an individually centered view of intelligence, most university graduates also take for granted this cognitive pattern ("structure" in Bateson's language) and thus give legitimacy to a misconception that J.B. Bury traces back to the Enlightenment thinkers who assumed that reliance on the rational process and experimental inquiry were cumulative and thus guaranteed progress into an infinite future (1932). Unfortunately, myth was seen as having been banished by the power of science, technology, and rational thought. Yet, there are many examples where myth continues to influence the development and use of modern technologies. For instance, the accumulated knowledge in the field of chemistry, which has led to natural systems (including the human body) being impacted by thousands of synthetic substances, such as DDT, PCBs, and dioxins, was all initially understood as break-throughs and celebrated as further examples of progress. Knowledge in other high-status fields of inquiry that most westerners associate with a linear form of progress are now being discovered to be ecologically problematic—yet the ways in which language perpetuates this recursive process continue to go largely unnoticed.

Bateson's reliance on the process of recursion to explain how we unknowingly perpetuate the misconceptions of the past leads to another basic insight that has particular importance for educational reformers. This insight was borrowed, as Bateson acknowledges, from Alfred Korzybski, a Polish-American philosopher and scientist. Bateson sums it up with the phrase "the map is not the territory."

The Map Is Not the Territory: How the Metaphorical Nature of Language Misrepresents the Differences Which Make a Difference

Bateson recalled the time when he arrived at the insight that enabled him to make the connection between Korzybski's distinction between map and territory and the epistemological issues he was working through. It was in 1970 while he was preparing his talk for the Korzybski Memorial Lecture. In response to the question he asked himself, "What gets from the territory onto the map?" the answer became clear that, as he put it, "News of difference is what gets across, and nothing else" (1991, p. 188). The connection Bateson made between Korzybski's now famous phrase and his own insight about what represents the most basic unit of information that undergoes constant transformation while circulating through all levels of the earth's ecosystems may appear quite mystifying.

Bateson was attempting to resolve the problem of the relationship between the mind and the external world—a problem that is ignored by nearly all public school teachers and most university professors even though their task is to provide the conceptual frameworks that will guide how students think about the external world, as well as their internal world. In elaborating further on the mind/external-world relationships, Bateson makes a further observation that is fundamental to recognizing the conceptual error that dominates education in the West. Following a restatement that what gets from the outside world into the brain is *"news of difference,"* he goes on to make the following observation: "If there is no difference in the territory, there will be nothing to say on the map, which will remain blank. And, further, I saw that any given map has rules about what differences in the territory shall be reported on the map." (1991, p. 200)

Map and territory are metaphors. The "map" refers to the cultural/metaphorical language/thought connections, while the "territory" is the world of the natural and cultural systems we commonly refer to as the environment in which we live. An especially important part of Bateson's statement about maps (a culture's way of knowing) is that they contain the interpretative frameworks that govern which differences which make a difference will be recognized and how they will be understood. For example, if only the increase in profits is given attention when growing genetically modified seeds, the differences which make a difference that signal environmental damage will go unnoticed. Similarly, focusing only on the loss of employment may lead to ignoring the carbon dioxide an industry is releasing into the atmosphere, which is contributing to changes in the ocean's chemistry. What Bateson is getting at in this statement about how cultural rules influence what will be recognized and how they will be interpreted is the role that the deep assumptions of the culture (root metaphors) play in framing what we are aware of as well as what will be ignored. To reiterate an important point: These cultural assumptions are largely taken for granted, thus leading to the process of selective awareness and interpretation that will be experienced as a natural, rather than as a cultural, construction.

The question that is likely to arise is: What does Bateson mean by saying that the differences which make a difference are the basic units of information, and why does he suggest that when differences are not present we have nothing to respond to? First, we need to clarify what he means by the statement that "a 'bit' of information is definable as a difference that makes a difference. Such a difference, as it travels and undergoes successive transformation in a circuit, is an elementary idea." (1972, p. 315) What he does not mean is that the "elementary idea" is like the metaphorical representation of ideas. Rather, it is the information that is processed at various levels—genetic, chemical, energy, behavior, thinking, etc.—that leads to a change in the organism or system that can process the way in which information is coded. For example, the introduction of toxic chemicals during the development of the fetus may represent a difference which makes a difference in terms of chromosomal damage that becomes, in turn, a difference which makes a difference in the development of the immune system—which then leads to a chain of differences that results in a variety of physical problems that then become a lifetime of differences which make a difference for the child, parents, and various social agencies. In this example the information communicated through differences circulates through interdependent systems where differences in the chemistry of an industrial process lead to differences in the functions of genes—and eventually to differences at the cultural level.

The introduction of a non-native plant starts another cycle of differences which make a difference. That is, when the chemistry of insects, including pollinators, does not fit with the chemistry of the non-native plant, differences which make a difference circulate through the local ecology—affecting the native plants that rely upon the bees and other insects essential to the pollination process, as well as the birds that rely upon the insects and the other participants in the local food chain. Bateson's seemingly simple phrase encompasses the information exchanges that support the self-development processes of every organism in an ecosystem as well as provide the sources of energy that are shared. When the different sources of energy are lost or changed, another complex set of differences circulates through the interdependent systems. Bateson is challenging the western idea that only humans are intelligent and can process information when he states that differences are sources of information and "an elementary idea." Ecosystems also process information in ways that are often beyond what humans can understand or replicate. What the dominant western epistemology fails to understand is that the natural environment is not reducible to matter and blind forces. Rather, it is sustained by the different ways in which "differences" are processed, and this is dependent upon eons of genetic development within different environmental contexts.

For Bateson, the maps are the metaphorical constructions that provide the interpretive and moral frameworks of the culture. What is important about the map/territory metaphor is that the map is rooted in an earlier cognitive/mythopoetic history. That is, the meanings of words (metaphors) are framed by people who are successful in having the analog they selected accepted by others, and even by later generations. Furthermore, their process of analogic thinking is framed by the taken-for-granted root metaphors of earlier times. In terms of the West, these root metaphors include patriarchy, anthropocentrism, mechanism, progress, individualism, economism, and now evolution—with ecology becoming a new totalizing interpretive framework which is challenging the earlier root metaphors that underlie the industrial/consumer-oriented culture. These root metaphors, or "cognitive structures," are the recursive epistemologies that are taken for granted today. They also underlie the process of linguistic colonization as Westerners attempt to force other cultures to base their daily lives on these root metaphors.

To state the problem more directly:

1   The metaphorical maps are generally out of date—that is, the long-established metaphorical language relied upon to respond to changes circulating through both the natural and cultural message systems contributes to the lack of recognition of what is most critical to slowing the rate of environmental degradation and the increase in social injustice.

2   The modern western-influenced maps, when relied upon by cultures existing in different bioregions, distort awareness of the interdependence between the local culture and the natural systems that needs to be taken into account if the rate of environmental degradation is to be reduced. For example, the cultures in the Peruvian and Bolivian Andes are unlikely to reduce the catastrophic consequences that lie immediately ahead (where the key difference which makes a difference is the melting of the glaciers that are the source of their water) if they rely upon the conceptual maps borrowed from the western cultures for guiding how they are to live—as these maps are major contributors to global warming.

3   The root and image metaphors need to be continually revised in order to limit even greater human suffering and environmental damage. Updating the maps requires being aware of the differences in the first place. This is not the old problem of what came first, the chicken or the egg—especially when it is understood that updating the maps upon which the next generation relies requires that classroom teachers and university professors be aware that there is an ecological crisis. They also need to recognize that science and technology alone will not help to mitigate the crisis, and that there are cultural traditions (recursive patterns) that are deepening the crisis. Solutions other than technological fixes are especially needed and will introduce students to community-centered and mutually supportive lifestyles that are less dependent upon consumerism and the industrial model that is based on a money economy and the pursuit of profits.

The Problem of Double Bind Thinking

The issues discussed up to this point—recursive looping where the present repeats the conceptual patterns and errors of the past, and the problem of conceptual maps that are outdated and of the wrong territories—represent different aspects of double bind thinking. Many of Bateson's original comments on the nature of double bind thinking were intended to clarify the abnormal communicative patterns of schizophrenics, as well as communication problems in general. However, what is most pertinent here are his views on how double bind thinking prevents us from moving beyond the recursive pull of basic misunderstandings of human/nature interdependencies. Peter Harries-Jones quotes how Bateson understood the connections of double bind thinking and the ecological crisis: "All communicative activity should be considered as a set of propositions about the world or the self, whose validity depends on the subject's belief in them. It [is] these beliefs *about* the world that should be the major topic of investigation." Harries-Jones further notes that "double bind, in Bateson's view, was never a matter of simple intellectual confusion or of being caught in a dilemma of 'I am damned if I do and I am damned if I don't.'" The double bind, for Bateson, involves "a situation in which simple dilemmas [are] compounded by falsified contexts, supported by patterns of interpersonal communication which ensured continuation of the denial that a falsified context [exists]." (1995, p. 135)

A falsified context can take many forms, such as the lack of awareness of the cultural construction of different interpretations of reality. Apathy and indifference toward exposing the reifications that lead people to take for granted that the interpretations are accurate representations of "reality" are yet other examples of a falsified context. The most prevalent examples of a falsified context involve relying upon a system of knowing borrowed from the distant past and used as a guide for understanding today's world, and thus representing language as being free of historical and cultural influences. Again quoting Harries-Jones, Bateson associates double bind thinking "with some combination of denial and inflexibility derived from the cultural predisposition about the salience of rationality and rejection of holism." Bateson is very specific about the nature of this inflexibility when he notes in one of his letters that "as long as the West remains tormented by a false pride in individualism, it will pursue perversions of individualistic thinking. This tormented perspective," he continues, "can lead to strategies in which killing the whole biosphere becomes preferable to risking one's own skin." (Harries-Jones, 1995, p. 227)

The Unit of Intelligence Is the Individual Plus the Immediate Differences Which Make a Difference— Plus the Recursive Cultural Epistemology that the Individual Takes for Granted

Bateson's challenge to the modern idea of individual intelligence has many dimensions. Included in the earlier quotation in which he refers to the "total self-corrective unit" are three key points:

1   Thinking, or what is generally referred to as the exercise of intelligence, is not like a process that leads to an idea that is like a photograph—such

as a mental image in which the boundaries are clearly framed off from the local contexts. Rather, Bateson refers to processing information in a way that involves continual self-correction and adjustments as differences which make a difference are taken into account.

2   Processing the elementary (and non-metaphorical) idea or information undergoing transformation as it circulates through the system occurs even when the individual is not involved. As Bateson puts it, "there are multiple differences between the *thinking system* and the 'self' as popularly conceived" (1972, p. 319, italics added). That is, the metaphor of intelligence needs to be expanded in ways that take account of the different ways in which information is coded and intergenerationally passed along.

3   The way the individual processes information is not always aligned with the cognitive/moral epistemology she/he inherits from the past. The individual may engage in self-correcting behavioral and reflective responses, while still taking for granted the reified ideas of an earlier time. Awareness of gender bias seldom involves recognizing the reified ideas of linear progress and a human-centered world. That is, the individual's reified ideas continue to perpetuate the state of self-denial—even as behaviors are undergoing changes in response to the differences circulating through the different pathways of information exchange that make up the larger ecology.

The question may arise about the classroom teachers and university professors who reinforce the idea that students are autonomous thinkers who are responsible for constructing their own ideas. This view of human intelligence carries with it the moral obligation in the West of citing from whom ideas are borrowed—or being deemed guilty of plagiarism. Or, is Bateson correct when he claims that the unit of intelligence (which he treats as a verb) is the individual, plus information circulating through the natural systems, plus the cultural epistemology that was constituted in the past and encoded in the languaging processes, plus the intergenerational legacy acquired as part of the taken-for-granted stock of knowledge? In *Steps to an Ecology of Mind,* Bateson uses the following example to highlight what is missing in the western view of individual intelligence. The example also highlights the interconnections between intelligence and the difference which makes a difference.

Consider a man felling a tree with an axe. Each stroke of the axe is modified or corrected, according to the shape of the cut face of the tree left by the previous stroke. The self-corrective (i.e., mental) process is brought about by a total system, tree-eyes-brain-muscles-axe-stroke-tree; and it is this total system that has the characteristics of immanent mind.

(1972, p. 317)

Other examples where the exercise of intelligence can be seen as participatory—and including the information flowing through the life-sustaining pathways of the larger system within which the individual is embedded—can be seen in the processes of non-verbal communication. A change in tone of voice, facial expression, or even a lengthy pause in a conversation, leads the person who is aware of the differences which make a difference to alter both behavior and thinking—which, in turn, leads to altering the response of the other person. Indeed, the adjustments take account of the response to what was previously said as well as the behavioral cues that accompanied what was said. The traditional farmer who is making a decision about when and where to plant a crop also exercises intelligence in a way that is influenced by the information circulating through the different interacting ecosystems of soil, plants, weather, quality of seeds from last year's harvest, and so forth. If the reader has had experience in sailing a boat, she/he will recognize that the decisions about the adjustment of the sail and rudder are continually modified in terms of the differences which make a difference in the water/wind ecology. Changes in the color of the water often signal a change in the velocity of the wind. The size of the waves and the direction of the current also influence the degree of heeling of the boat, and the change in tack always takes account of the direction of the wind as well as where one hopes to arrive. Life in the natural environment is also affected by differences which make a difference. Gary Snyder puts it this way: "The world is watching: One cannot walk through a meadow or forest without a ripple of report spreading out from one's passage. The thrush darts back, the jay squalls, a beetle scuttles under the grasses, and the signal is passed along. Every creature knows when a hawk is cruising or a human is strolling. The information passed through the system is intelligence." (1990, p. 19)

All of the above examples involve giving careful attention to relationships. Changes in the cut face of the tree, the non-verbal patterns of communication of the person with whom one is engaged in conversation, the soil that is to be planted, and the course that is to be sailed, could be (and too often are) erroneously thought of as separate things, entities, and objectives. When the interactive relationships are ignored, the exercise of intelligence becomes formulaic and a preconceived strategy is put into play. When this occurs, the information circulating within the natural and cultural systems becomes ignored, with attention being given to what the individual has been culturally conditioned to be aware. That is, the old conceptual maps take over, with the individual's awareness being limited primarily to what the misconceptions of earlier thinkers bring into focus. For example, today's market liberals, whose focus on achieving greater profits is guided by the abstract theories of classical liberal thinkers, do not consider the differences which make a difference in the cultural and natural ecologies in which they are embedded. This leads them to ignore the differences introduced by their actions, such as increased levels of poverty, deskilling of workers, increases in toxic pollution, damage to the self-renewal of natural systems, and so forth. That is, if they were educated in a manner that reinforced the importance of giving attention to relationships—rather than rigidly being guided by the abstract free-market ideology—perhaps they would recognize another point that Bateson makes. Namely, that "in no system which shows mental characteristics can any part have unilateral control over the whole. In other words, *the mental characteristics of the system are immanent, not in some part, but in the system as a whole.*" (1972, p. 316)

This statement relates directly to the moral values that should be integral to the exercise of ecological intelligence. The epistemological shift from focusing on things to relationships also involves a shift in the role that language plays in carrying forward the culture's moral templates. Metaphorical thinking, which is framed by the analogs settled upon by earlier thinkers, carries forward how they understood the attributes of things, such as trees, wilderness, the ocean and rivers, non-native plants, animals, and so forth. For example, when wilderness was understood as a source of danger, it was both rational and moral to treat it as an exploitable resource. Similarly, plants not considered to have any useful attributes were called weeds and in need of being eradicated. One of the attributes of the oceans, namely their vastness, led to thinking of them as impervious to human impact and thus moral responsibility. Because insects were thought to be lacking in useful attributes, exterminating them with a pesticide was a morally appropriate behavior. Root metaphors, such as anthropocentrism and progress, provided moral legitimacy for introducing into the environment thousands of synthetic chemicals that we are only now recognizing as part of the emerging health catastrophe that is the legacy of early and current scientists. The root metaphor that represented the world as a collection of things, which included autonomous individuals, framed how earlier thinkers understood the attributes of things that ranged from women, indigenous peoples, pre-literate cultures, and so forth. Reducing them to things rather than recognizing their relationships within their cultural and natural ecologies, which would have led to a more complex understanding, made it easier to label each as possessing only a negative attribute—which, in turn, made it unnecessary to be morally accountable toward them.

Bateson's emphasis on understanding ecosystems as layered, interactive, and interdependent self-renewing systems, ranging from genes to cultural assumptions, leads to a shift in how moral values are to be understood. He recognizes that in some systems the relationships are disruptive and thus are ecologies that are not likely to survive. He refers to them as an ecology of weeds and bad ideas. One of these bad ideas is that humans, by relying upon the rational process and new technologies, will be able to survive the destruction of natural systems. In a passage that recalls his criticism of the West's recursive epistemology which continues to separate the fate of humans from the fate of the environment, he issues the following warning:

The environment will seem to be yours to exploit. Your survival unit will be you and your folks or conspecifics against the environment of other social units, other races, and the brutes and vegetables. If this is your estimate of your relation to nature *and you have an advanced technology,* your likelihood of survival will be that of a snowball in hell. You will die either of the toxic by-products of your own hate, or, simply, of over-population and overgrazing. The raw materials of the world are finite.

(1972, p. 462)

Following this passage, Bateson goes on to say that the most important task today is to learn to think in a new way. Before considering what he describes as the three levels of learning, and how the latter level leads to what can be called ecological intelligence, it would be useful to address a response that both philosophers and educational theorists are likely to make. Because Bateson appears at first glance to be a process thinker, they are likely to associate his ideas with those of John Dewey. This would be a major mistake, and for the following reasons.

Basic Differences Between the Ideas of Gregory Bateson and John Dewey

On the surface there appear to be many similarities between Bateson and Dewey. Both understand that knowledge has to be continually revised in order to take account of a constantly changing world. A second surface similarity is that both reject the idea that intelligence is an attribute of the autonomous individual. For Dewey, intelligence involves problem solving in a democratic context which becomes more efficient as communication with others is enhanced. The argument that Dewey was an early environmental thinker, which would suggest another favorable comparison with Bateson, is based on interpreting Dewey's understanding of intelligence as an integral part of experience—and experience as part of the natural world. This view of intelligence avoids the error inherent in the Cartesian mind/body separation of which Bateson is also critical.

Given these surface similarities between Dewey and Bateson, educational reformers who have recently recognized that there is an ecological crisis, and who are searching for a conceptual framework that will guide their thinking, are likely to feel that their years of relying upon Dewey's progressive and democratically oriented theory of knowledge make it unnecessary to take on the challenge of understanding Bateson's admittedly difficult vocabulary and concepts. However, if these reformers were to examine the differences in any depth, they will recognize that Dewey, for all of his useful insights, is part of the problem. Let me cite the following as evidence. First, Dewey grew up during successive waves of environmental devastation, such as the killing off of millions of bison, the clear-cutting of forests across the country, the destruction of prairie grasses—not to mention his support of the industrial processes that were spewing billions of tons of carbon dioxide and other toxic chemicals into the rivers and into the atmosphere. He says nothing about the environmental destruction of his era. In fact, while he wanted democratic socialism to replace capitalism, he also thought that the growth and successes of the industrial culture would lead to wider acceptance of the scientific and experimentally oriented approach to knowledge.

Second, during his most formative intellectual years, the indigenous cultures were being decimated, by some estimates, to 90 percent of their previous population. Their lands were being taken over by the Anglo/Euro-Americans, and Dewey remained silent. His understanding of the indigenous cultures, which exhibited many of the characteristics of ecological intelligence, is summed up in several books in which he describes them as having the thought patterns of "savages." Dewey's racism has been defended on the grounds that he shared many of the taken-for-granted prejudices of his era. This seems a weak excuse, especially in today's world when there is an increasing awareness of the connections between linguistic diversity and preserving biodiversity. There is another aspect of Dewey's thinking about other cultural ways of knowing, which he lumps together under the category of "spectator knowledge," that makes his theory of knowledge and the educational reforms derived from it especially problematic. He does not represent instrumental experimental inquiry as just one of many approaches to knowledge. Rather, it is the only legitimate approach. Dewey's colonizing mentality leads to reducing all forms of knowledge to three categories: savage, spectator, and experimental inquiry. These categories represent Dewey's way of understanding the stages of social progress. He was, like other intellectuals of his era, a Social Darwinian thinker who was driven by the idea that if the educational process teaches students the importance of participatory decision-making in solving problems by using the scientific mode of experimental inquiry, they will be able to escape the intellectual prisons of their immigrant parents. For Dewey, there is only one valid approach to knowledge, and this approach requires overturning the traditions of intergenerational knowledge that sustains the cultural commons of these diverse immigrant groups.

A criticism that can be made of Dewey, which is the same one that Bateson makes of scientists, is that Dewey was not a reflexive thinker. His assumptions about the progressive nature of experimental and participatory problem solving led him to ignore the deep cultural assumptions that led to his silence about the environmental devastation and the threat that the industrial model of production and consumption posed for the environment and other cultures, as well as the ecological knowledge of the indigenous cultures he labeled as savages. While Edward Sapir and Benjamin Lee Whorf were beginning to explore the connections between language, ways of knowing, and cultural practices, Dewey remained indifferent to the reality constituting role of language—particularly how the metaphorical thinking of earlier eras carries forward their misconceptions and silences. Friedrich Nietzsche was writing about this problem in the 1880s, so it would be unfair to excuse Dewey for being unaware of the cultural/metaphorical language issues that are receiving such wide attention today. The important point is that today's followers of Dewey reproduce in their own thinking about educational reforms the same silences that resulted from Dewey's lack of reflexive thinking.

Bateson avoids adopting any of the prejudices that characterize Dewey's thinking. Indeed, it is difficult to find any reference to Dewey in Bateson's writings, just as it is difficult to find in Dewey's writings any reference to ecology—even though the word was widely used in the early 1900s to refer to the study of natural systems. My suggestion, in light of the rate of changes taking place in the earth's ecosystems, is that thinking about the educational reforms that contribute to an ecologically sustainable future should focus on developing a deeper understanding of ecological intelligence—including Bateson's contribution to understanding the double binds that inhibit the educational reforms that foster ecological intelligence. Dewey can be credited with introducing educational reforms during an era of rote learning and childhood repression, but these reforms are now widely accepted. It is time for his followers to begin addressing reforms that foster lifestyles and patterns of thinking that are less damaging to the environment. Among these reforms are lifestyle changes that do not fit with Dewey's emphasis on continual change and experimentation—which he associated with progress in moving beyond the non-scientifically grounded traditions of the past.

Exercising Ecological Intelligence and Level III Learning

A good place for starting this transition from the recursive epistemology of ecologically problematic ideas is by following Bateson's suggestion that we need to move beyond what he calls Learning I and II, by participating in Learning III. In an essay on learning written in 1964 and revised in 1971 for inclusion in *Steps to an Ecology of Mind*, Bateson summarizes the scientific research on Learning I and II. Learning I is limited to responding to a stimulus and then correcting the choice being made when given a set of alternatives. This is the form of learning observed in studies of rat behavior. Learning II involves a more complex set of responses, such as being aware of changes in the context within which choices are made. It also includes a range of attitudes that influence the process of learning. These include being fatalistic (e.g., accepting a given set of relationships and possibilities), an inability to question the otherwise tacit understandings of relationships and contexts, adopting an attitude of dominance or submissiveness that closes off recognition of other possible relationships and ways of thinking, and adopting a pattern of thinking where events are understood to be discrete rather than interconnected. To this list can be added learning within the limits established by reified beliefs and traditions.

If we translate the list of characteristics associated with Learning II into more contemporary language, it then can be understood as the ability to learn in contexts dominated by the expectation that events are beyond human intervention, that the culture's beliefs and values are taken for granted (which means that their cultural origins will not be recognized), that one's sense of authority and right to dominate others is absolute (either derived from God or a reified ideology), and that events are to be judged without consideration of their antecedents or future consequences. This level of learning is likely to ignore that others may have different interpretations and even different belief systems. Other characteristics include a willingness to accept the authority of the printed word and abstract knowledge, especially when they help to give legitimacy to ideas and values that the individual claims to originate. In short, Learning II can be seen in the cognitive style of the authoritarian personality. It can also be seen in the cognitive style of the indifferent and passive individual who seeks strength in following social conventions—even those that serve the interests of authoritarian individuals. Both types, and the many individuals who are both authoritarian in some areas and who find strength in belonging to emotionally charged mass movements, view themselves as individuals who are not dependent upon either culture or the natural environment. Their sense of autonomy leads to their thinking that they have no responsibilities except for what serves their personal interests or that of their immediate family.

That people engaged in Learning II are not the only ones existing in society led Bateson to consider the characteristics of people who exhibit Learning III characteristics. These are the characteristics that are essential for moving from an individually centered intelligence (Learning II) to that of ecological intelligence. Among the qualities Bateson associates with Learning III are:

1   An ability to question the premises underlying both one's own behavior as well as practices and policies that govern society;
2   A willingness and conceptual ability to question what is taken for granted by Learning II individuals, and to introduce changes;
3   An awareness of the importance of understanding differences in cultural contexts;
4   An ability to assess habits (whether personal or culturally shared) in terms of whether they need to be revised, changed completely, or conserved, such as conserving habeas corpus and other civil liberties, as well as those aspects of the cultural commons that reduce dependence upon a market economy; and
5   An awareness of cultural continuities and interdependencies in both cultural and natural ecologies—and of the conceptual double binds that put the well-being of both at risk.

These characteristics are mutually supportive, and if taken seriously would lead to profound reforms in both public schools and universities. Bateson recognizes the difficulty in making the transition to Learning III, so the question might come up as to why we should persist in recommending reforms in the two institutions that most people operating at the Learning III level regard as tradition-bound—even as these institutions appear on the cutting edge of promoting even more extreme forms of modernism. (I use the word "tradition-bound," instead of "conservative," as the latter is chronically misused in today's political discourse. Throughout the book, the argument is made that the exercise of mindful conservatism and ecological intelligence are essential to maintaining the cultural commons.)

If public schools and universities are continuing to reinforce the same deep cultural patterns of thinking that gave conceptual direction to the consumer/industrial culture that is now being globalized, even as some professors are working on new technological solutions, why argue that attention should be focused on reforming the modernizing traditions of public schools and universities? If we keep in mind that one of the principal characteristics of Learning III is the ability to question the premises upon which the taken-for-granted cultural practices are based, it quickly becomes obvious that there are few institutions in the mainstream culture where the underlying premises (what I prefer to call "cultural assumptions" and "root metaphors") can be questioned without facing personal, economic, and political consequences. Many families will not allow the assumptions that guide their economic, political, and moral decisions to be questioned and revised, and there are few other social organizations and institutions that would welcome this sort of questioning. The list includes most churches, community organizations such as the local city club and chamber of commerce, the Rotary Club and other civic organizations, National Guard gatherings, local sailing clubs and other sporting groups, gun clubs, places of employment—indeed, the list goes on and on.

I have serious reservations about whether most classroom teachers and university professors possess the conceptual background necessary for recognizing why many of the taken-for-granted cultural assumptions that underlie their academic discipline are ecologically problematic, or the willingness to take the ecological crisis seriously enough to begin questioning these assumptions. Nevertheless, public schools and universities are the two institutions that provide what can be called the psychosocial moratorium necessary for raising difficult questions and obtaining an historical perspective on how, in the name of progress, intellectual elites have succeeded in poisoning much of the environment, and in promoting a form of individualism that equates a level of consumerism that is ecologically unsustainable with personal happiness and success. While public schools have no tradition of academic freedom, they nevertheless can provide students with the initial conceptual basis for making the transition to exercising ecological intelligence without embroiling the school in controversy. The tradition of academic freedom is well established for universities, which allows for a more far-reaching examination of the guiding assumptions of the dominant culture. Thus, universities have the fullest potential for providing the conceptual space necessary for students to move to Learning III and to exercising ecological intelligence.

The task in the following chapters is to examine the differences between individual and ecological intelligence for how we think about social justice issues, the prospects of democracy, and the moral values that will guide our relationships as we enter an era of scarcity of water, protein, and habitable land. We will then turn to consider the current traditions in teacher education that must be taken into account before taking on the challenge of identifying the educational reforms that can be put into practice. This will involve examining how the educational uses of computers promote abstract thinking while marginalizing awareness of local contexts and the differences which make a difference in cultural and natural systems. Also to be considered is how the education section of UNESCO is promoting reforms in teacher education that are based on the double bind thinking of equating critical thinking and the individual construction of knowledge with educating for sustainable development. The last chapter will focus on how the Quechua of the Peruvian Andes have maintained their traditional practices of ecological intelligence.

References

Bateson, Gregory. 1936. *Naven: A Survey of the Problems Suggested by a Composite Picture of the Culture of a New Guinea Tribe.* Cambridge, G.B.: University of Cambridge Press.

Bateson, Gregory. 1972. *Steps to an Ecology of Mind.* New York: Ballantine Books.

Bateson, Gregory. 1979. *Mind and Nature: A Necessary Unity.* New York: Bantam Books.

Bateson, Gregory. 1991. *A Sacred Unity: Further Steps to an Ecology of Mind.* New York: A Cornelia & Michael Bessie Book.

Bateson, Gregory, and Mary Catherine Bateson. 1987. *Angels Fear: Towards an Epistemology of the Sacred.* New York: Macmillan Publishing Co.

Bury, J.B. 1932. *The Idea of Progress: An Inquiry into its Growth and Origins.* New York: Dover Publications.

Goleman, Daniel. 2009. *Ecological Intelligence: How Knowing the Hidden Impacts of What We Buy Can Change Everything.* New York: Broadway Books.

Harries-Jones, Peter. 1995. *A Recursive Vision: Ecological Understanding and Gregory Bateson.* Toronto: University of Toronto Press.

Havelock, Eric A. 1986. *The Muse Learns to Write: Reflections on Orality and Literacy from Antiquity to the Present.* New Haven, Conn.: Yale University Press.

Lakoff, George, and Mark Johnson. 1999. *Philosophy in the Flesh: The Embodied Mind and its Challenge to Western Thought.* New York: Basic Books.

Rand, Ayn. 1964. *The Virtue of Selfishness: A New Concept of Egoism.* New York: Penguin.

Snyder, Gary. 1990. *The Practice of the Wild.* San Francisco, Calif.: North Point Press.

12 Rethinking Social Justice Issues Within an Eco-Justice Conceptual and Moral Framework

As the social justice issues of class, race, and gender have been the dominant concern of many teacher education and educational studies faculty over the last decades, it is now time to ask whether the recent evidence of global warming, changes in the chemistry of the world's oceans, and the increasing shortage of potable water should lead to developing a new strategy for ameliorating these long standing sources of injustice, homelessness, and poverty. Given the amount of time devoted to discussing class, race, and gender issues with students, as well as the number of books that focus on these issues, little has actually been achieved in effecting the systemic changes required for marginalized social groups to participate on more equal terms in the public arenas of politics, economics, and educational opportunities. Corporations in the United States continue to shape governmental policies that deepen the economic plight of marginalized groups who live at the bottom of the wage scale, while raising the cost of drugs and medical care beyond what they can afford. Overall, the democratic process itself has become degraded by corporate and other special interests to the point where millions of people continue to be mired in poverty and hopelessness.

The recent acceleration of economic globalization, and the deepening of the ecological crises that are now impacting people's daily lives, suggests that a radical rethinking of how to address social justice issues is needed. The growing awareness of these global developments, which includes the lack of moral constraints on the free market system, a weakened labor movement and rapid rise in unemployment, the decline in the size of the middle class, and a need to change the ecological impact on all citizens (even that of the poor who have not been educated about how to live less environmentally destructive lives), means that the old assumptions about achieving a more socially just society need to be re-examined.

Social justice thinking has largely been framed in terms of middle-class assumptions about individualism, progress, a world of unlimited exploitable natural resources, and education as a source of individual empowerment. The ultimate goal of achieving greater social justice for marginalized groups has been to enable them to participate on equal terms in the areas of work, politics, and the culture of consumerism. The guiding priorities of ecojustice-based educational reforms are, on the other hand, both more global in terms of analysis and accountability, and more local in terms of educational strategies that reverse the process of deskilling that is part of the destruction of community systems of mutual support that began with the rise of the techno-scientific based industrial culture. These priorities can be summarized as eliminating environmental racism, resisting the forces that are colonizing Third World cultures and exploiting their natural resources, revitalizing the local cultural and environmental commons that are sites of resistance to the expansion of the industrial/consumer dependent lifestyle, adopting a lifestyle that does not degrade the prospects of future generations, and developing an ecological consciousness that respects the right of natural systems to renew themselves.

Basing daily life on ecojustice criteria means taking account of the impact of the consumer-dependent lifestyle that is being promoted in our public schools and universities by asking whether it is largely responsible for the economic and cultural colonization of Third World societies, as well as the environmental racism that exposes minority groups to the toxic chemicals that the industrial/consumer-oriented culture relies upon. Ecojustice thinking also brings into focus the need to consider the existing community-centered alternatives to the deskilled individual lifestyle that is increasingly dependent upon consumerism—even as the sources of employment become more uncertain because of outsourcing to low-wage regions of the world, and the drive to increase profits by replacing workers with computer-driven machines. As the life-sustaining ecosystems become more degraded, there is also the question of whether the current industrial/consumer-oriented lifestyle that is taken for granted by many educational advocates of social justice is undermining the prospects of future generations. Other concerns of ecojustice thinking include the need to undertake educational reforms that address our responsibility for leaving future generations with sustainable ecosystems, which also means recognizing the right of non-human forms of life to reproduce themselves in sustainable ways.

While the environment is being degraded to the point where the scarcity of protein, water, and energy is driving up prices, thus further impoverishing the already poor, the advertising industry is spending billions of dollars a year in order to perpetuate the public's addiction to consuming the latest fashions, technologies, and forms of entertainment. Public awareness of the environmental changes that scientists are warning about is further obfuscated by the big-box stores and shopping malls that stock their shelves with a super abundance of consumer products—thus further perpetuating the illusion of plenitude. Glitz, easy credit, and continued indifference to the dangers of going deep into debt are just part of the culture that now dominates the majority of the people's lives—that is, those who have not lost their well paying jobs, health and retirement benefits, and are not now reduced to a minimum-wage and near subsistence lifestyle. The poor and marginalized—ranging from single mothers, urban minority youth, migrant farm workers, and a wide range of people whose skin color and lack of educational back-ground disqualify them from other than menial forms of labor in industrial food outlets and other low-paying service-industry jobs, are too focused on meeting the most basic needs of food and shelter to be aware that there are community-centered alternatives to the industrial/consumer lifestyle they have been excluded from participating in. As Barbara Ehrenreich pointed out in an interview with Bill Moyers (2007), the poor live so close to the edge that going without pay for the couple of weeks it takes to find a more high-paying job is unthinkable. In effect, poverty restricts even this most basic option that the middle-class can take for granted. There are now millions of people who are unable to find jobs of any kind, which is a problem that educational reformers have not yet addressed.

The central priorities of ecojustice advocates do not have their roots in abstract theory. Rather, the traditions of intergenerational knowledge and patterns of mutual support that enable people to live in ways where market forces do not dominate everyday life have been around since the beginning of human history. They are still present in every community across North America and in other parts of the world. Historically, these traditions were known as the commons; that is, what is freely shared by the members of the community—which also includes local decision-making. The norms that governed the cultural and environmental commons were passed along orally and differed from culture to culture. The Romans were the first to establish a written record of the commons, which they identified as the local streams, woods, fields, animals, and so forth. The cultural commons, which include the intergenerational knowledge and skills necessary for gathering, preparing, and sharing food, the medicinal properties of plants and where to find them, narratives of courage and of hubris, the rules that governed community members who violated local norms of justice, the sharing of technological skills and craft knowledge, the mythologies and prejudices that regulated who had privileged positions in the community, and so forth, have only recently been identified as part of the commons. The cultural commons also include the voluntary associations that are sources of mutual support within the community, as well as groups that come together to promote agendas that range from providing safe bicycle lanes within the commu-nity to supporting the peace efforts of national politicians, and providing aid to people in other parts of the world that have experienced a natural disaster. Unfortunately, the intergenerational sources of empowerment and community self-sufficiency are now being threatened by the market-liberal traditions of private property, anomic individuals who have made a virtue of their cultural amnesia, the expansion of the industrial approach to produc-tion and consumption, the growing hegemony of the capitalist ethos, and the rise of corporate power.

While the causes of the economic crisis that is spreading around the world are systemic as well as a function of human greed, it is important to note that the main focus of the media, social-justice politicians, and public is on regaining the jobs that will enable people to return to their previous consumer-dependent lifestyles. That is, there is little discussion of community-centered alternatives that are here being referred to as the cultural commons—and thus little discussion of how the political economy of the local cultural commons can become part of the basis for meeting the daily needs for food, housing, medical care, and education. If attention were to be given to the lifestyles of people who are more fully engaged in their local cultural commons it would become clearer that they rely less upon a money economy and thus are less exposed to the exploitive forces that are inherent in the industrial governed market place. The voluntary simplicity movement has demonstrated that the political economy of the cultural and environmental commons leads to a different understanding of wealth—one that takes account of skills, mutual support systems, and community well-being. Yet, it is important to recognize that there is still a need for meaningful forms of work that contribute to a living wage. The combination of local decision-making, which is a key feature of many local cultural commons, and the spread of such developments as micro-banks and the pooling of local resources for housing projects and other community infrastructure needs, are also evidence of the need to combine thinking about local self-sufficiency and ecological sustainability.

The relationships between the local cultural commons found in every community today and the industrial/consumer culture have not been mutually supportive. Indeed, the people who promote the expansion of the industrial/consumer-dependent lifestyle, and thus the accumulation of capital, view the largely non-monetized cultural commons as potential markets to be exploited. Their goal is to replace intergenerational skills and patterns of mutual support with new technologies that must be privately owned and with expert systems that represent as sources of backwardness the traditional values and forms of knowledge—such as civil liberties, patterns of returning labor, mentoring, and knowledge of how to live lightly on the land, that have been the strength of many cultural commons. At the time the environmental commons in rural England were being transformed during the early stages of the Industrial Revolution, the process of limiting free access and use on a non-monetized basis, as well as the overturning of local decision-making, was referred to as "enclosure". That is, the enclosure of the environmental commons involved the introduction of private ownership and integration into a money economy, which often led to decision-making being transferred to distant owners—and later to corporations that made increasing profits their primary goal.

Now that we can recognize the cultural beliefs and practices, which now include cyberspace, as part of the cultural commons that enable community members to be less dependent upon a money economy, it is possible to recognize the many ways in which different aspects of the local cultural commons are being enclosed by today's market forces—as well as by ideologies, technologies, prejudices, and silences. Public schools and universities continue to be complicit in reinforcing the cultural assumptions that further undermine the viability of the cultural commons, especially the cultural commons of ethnic groups, even as environmental scientists are working to conserve what remains of the environmental commons. Many social-justice oriented faculty continue to reinforce many of the same cultural assumptions that gave conceptual direction and moral legitimacy to the industrial/consumer-dependent lifestyle even as they criticize the exploitive nature of capitalism. These shared assumptions include the idea of the autonomous individual, the progressive nature of change, an anthropocentric view of human/nature relationships, and the drive to impose these assumptions on other cultures under the rubric of "development", as well as the same silences about the local community sources of self-sufficiency and mutual support. (Sachs, 1992).

While the diversity of the world's cultural commons currently represents sites of resistance to economic globalization, it is important to avoid romanticizing the cultural commons. In many cultures, including the local communities across North America, the cultural commons also include narratives and traditions that perpetuate different forms of discrimination and economic exploitation. That is, the stoning to death of the woman who seeks to marry outside of her tribe, the market-liberal ideology that equates social progress with an economy that makes survival of the fittest the ultimate test of individual success, and the various forms of racial, class, and gender prejudices also have their roots in the traditions of some cultural commons. Ironically, these non-monetized traditional beliefs and practices (which have dire economic and social consequences for those who are the subjects of discrimination) were and still are generally sustained in communities which may also possess networks of mutual support that reduce reliance on consumerism—and that have a smaller ecological footprint.

The local cultural commons should not be regenerated and supported just because they represent alternatives to the industrial/consumer-oriented culture that is being globalized. Rather, the different traditions of the cultural commons need to be examined in terms of whether they support traditions of civil liberties, as well as moral reciprocity in the treatment of all members of the community as deserving the right to an equal opportunity to develop their personal talents and to make their contributions to regenerating the life-supporting cultural commons. Challenging the traditions of the cultural commons that are sources of exploitation and marginalization should also be part of a more global and ecologically informed ecojustice pedagogy.

As pointed out in *Educational Reforms for the 21st Century: How to Introduce Ecologically Sustainable Reforms in Teacher Education and Curriculum Studies* (2011), the unique characteristics of the cultural and environmental commons require a radically different approach than the current emphasis on making individual emancipation, promoting the students' construction of their own knowledge, and making higher test scores the primary foci of educational reform. There are a number of unique characteristics of the cultural commons that an ecojustice pedagogy needs to take into account. The first is that most of the traditions that members of a community participate in on a daily basis are taken for granted, such as the tradition of English speakers using the subject-verb-object pattern of oral and written communication, assuming they are innocent until proven guilty before a jury of peers, and using language as a conduit in a sender/receiver process of communication—to cite just a few of the taken for granted patterns of daily life. The taken for granted status of most aspects of the local cultural commons is important for several reasons. In being part of tacit, contextual, and largely taken for granted experience, they are mostly excluded from the curriculum of public schools and universities. In being excluded from the curriculum at all levels of the formal education process, and in being largely taken for granted by members of the community who are at the same time being constantly indoctrinated with the message that change is essential to progress, the loss (that is, enclosure) of different traditions of the cultural commons too often go unnoticed. The exceptions are the people who are consciously carrying forward one of the traditions of the cultural commons, such as weaving, protecting civil liberties, utilizing craft knowledge and skill, participating in local theater, and so forth.

The taken for granted nature of the individual's experience of the cultural commons, which may include racist and gender forms of discrimination, is just one of the characteristics of daily experience that require a different approach to teaching and learning than is found in current approaches that are based on many of the same cultural assumptions (or what I have referred to elsewhere as root metaphors) that underlie the industrial/consumer-oriented culture that is overshooting the life-sustaining capacity of natural systems. The emphasis on explicit forms of knowledge, which is reinforced by reliance on print-based knowledge, testing, and supposedly objective knowledge, marginalizes the importance of helping students recognize the differences between their experiences in the cultural commons and in the market/consumer-oriented culture.

Another bias in current approaches to education can be traced back to Plato's argument that *pure thinking* leads to universal truths that are more reliable than thinking grounded in embodied/culturally-influenced experiences. The western theorists who followed in this tradition of assuming that abstract words are a more accurate source of knowledge also were unaware of the nature and ecological importance of their local cultural commons. Indeed, they held in contempt the forms of face-to-face, intergenerationally shared knowledge and skill, and relegated them to low-status knowledge.

This tradition is still evident in the thinking of current educational reformers who assume that words such as individualism, democracy, tradition (which reproduces the Enlightenment assumptions of being a source of backwardness and special privileges), intelligence, and progress have a universal meaning. (Ayers, Quine, and Stovall, 2009). These educational reformers continue to ignore how the analogs that frame the meaning of these metaphors carry forward the misconceptions of earlier thinkers. This pattern of thinking further marginalizes an awareness of the embodied experiences in the different community traditions that are being referred to here as part of the cultural commons. One of the consequences of the silences about the nature and complexity of the cultural commons, as well as the constant reminder that traditions are impediments to progress, which are being reinforced in most areas of the public school and university curriculum, is that students enter adulthood without an awareness of the different economic and ideological forces that are enclosing what remains of the cultural commons. For most of them, the industrial/consumer culture is the arena in which they will personally succeed or fail—and the outcome of their individual quests remain disconnected in their thinking from the rapid rate of degradation of the world's ecosystems.

There is now a major body of writing that addresses both the various ways in which public schools reproduce the culture's traditions of class and other forms of discrimination, as well as the reforms that need to be undertaken in order to achieve a more equitable society. Criticism of prejudicial language, silences in the curriculum, preconceptions about the potential (or lack thereof) of already marginalized students, tracking and other systemic forms of discrimination, have been the mainstays of educational foundations and educational studies courses for the past several decades. While there have been some social justice gains, particularly in the areas of race and gender, there remains much to be done—especially since the changes resulting from economic globalization and global warming will have the greatest impact on minority groups whose economic gains have been, at best, both minimal and fragile. Critiques of the beliefs and values that have kept people of color, women, and other people restricted by other class barriers have actually been critiques of the reactionary traditions found within some cultural commons. Unfortunately, the theories that framed these critiques were not informed about the complex nature of the cultural commons. The main consequence of this lack of understanding is that the aspects of the cultural commons that hold out the prospect of finding community-centered alternatives to the negative impact of the industrial culture have not been part of the well-intended efforts to use the schools to eliminate the sources of poverty and injustice.

The use of a sociological interpretative framework seemed ideally suited to bringing into focus economic, political, and educational inequities. Unfortunately, it has led to ignoring the questions that would have arisen if a more anthropologically informed interpretative framework had been relied upon. Awareness of a deep understanding of cultural differences could easily have brought into question how notions of individual freedom and equality could be reconciled with the importance that has been given in recent years to avoiding cultural colonization. For example, the western ideal of individual freedom and the diversity of non-western cultures do not easily fit together.

Another limitation of the sociological interpretative framework is that it keeps the analysis and recommendations for reform on human-to-human relationships, with the human-to-nature relationships being ignored. The evidence for this claim can easily be substantiated by reading educational writers who have most influenced how the analysis of class, race, and gender has been framed—especially writers such as Samuel Bowles, Herb Gintis, Michael Apple, Henry Giroux, and Peter McLaren. Recently, however, Bowles and Gintis have been writing about the importance of the commons, and McLaren has now turned attention to explaining how Marxism can guide educational reforms that address issues of sustainability. (McLaren, 2005) The key point is that today's educational discourse on class, race, and gender continues to ignore, with only a few exceptions, the implications of the ecological crises for the very social groups they want to emancipate.

The use of the cultural commons as the conceptual framework for analyzing the various forms of discrimination, as well as for guiding educational reforms, has several advantages that a sociological framework lacks. To reiterate: the cultural commons represent all of the forms of knowledge, values, practices, and relationships that have been handed down over generations that have been the basis of individual and community self-sufficiency—and that have enabled members of the community to be less dependent upon a money economy now undergoing systemic changes. While the previous discussion of the reactionary and, in some cases, horrific practices of some of the world's cultural commons need to be kept in mind, there are other characteristics of self-sufficiency that existed prior to what Karl Polanyi called the *Great Transformation,* when the emergence of the industrial system of production led to the enclosure of the environmental commons. (Polanyi, 2001) Kirkpatrick Sale summed up in *Rebels Against the Future: The Luddites and Their War on the Industrial Revolution* how the survival and global expansion of the industrial system of production and consumption depended upon the enclosure of the cultural commons. (1995) As he put it:

> All that "community" implies—self-sufficiency, mutual aid, morality in the market place, stubborn tradition, regulation by custom, organic knowledge instead of mechanistic science—had to be steadily and systematically disrupted and displaced. All the practices that kept the individual from being a consumer had to be done away with so that the cogs and wheels of an unfettered machine called the 'economy' could operate without interference, influenced merely by invisible hands and inevitable balances"
>
> p. 38.

Sale does not refer to the community traditions of self-sufficiency as the cultural commons, but he accurately makes the point that the industrial/consumer-dependent culture requires the destruction of the different forms of intergenerational knowledge, skills and mutually supportive relationships that enabled people to live less consumer-dependent lives. In effect, he is describing how the success of the industrial system of production and consumption required the destruction of the local cultural and environmental commons. What is ironic is that the kind of individual required by the industrial/consumer-dependent culture is the autonomous individual being promoted by many of today's educational reformers. It is also important to note that the lack of intergenerational knowledge that reduces dependence upon consumerism contributes to an important aspect of poverty that is seldom discussed—even though it leads to the forms of poverty that threaten the individual's health and leads to other forms of insecurity.

Unlike the limited conceptual possibilities of a sociological interpretative framework and vocabulary, the cultural commons is the phrase that encompasses the traditions of community that are nested in larger social and ecological systems. These traditions, as mentioned earlier, range from local approaches to growing and preparing food (as alternatives to an industrialized system) that is so damaging to ecosystems and to intergenerational approaches to healing that differ from the highly monetized and industrial approaches of today's medicine (which are increasingly becoming dependent upon patenting indigenous knowledge of the medicinal properties of plants). (Shiva, 1996) Depending upon the local community and cultural traditions, the intergenerational knowledge also includes the creative arts passed on through mentoring that differ from the star system of commercialized music and visual arts; and as well as narratives of the labor, feminist, and civil rights movements rather than the mind-numbing television sit-coms that also serve to hook viewers to the multi-billion advertising industry. The traditions of civil rights that go back to the Magna Carta of 1215 are also part of the cultural commons.

Unfortunately, they are now being enclosed by the growing alliance between market-liberal-dominated governments, corporations, universities, and the military establishment. A more fine-grained analysis of the differences between the cultural commons and the industrial/consumer-dependent culture that is now being globalized would involve a discussion of the differences between community mentors and university-trained experts who have an ego and economic investment in imposing abstract theory-based solutions on people's lives, between face-to-face and computer-mediated communication, between community traditions of reciprocity where work is returned and work that has to be paid for, between developing personal interests and skills and being a consumer of other people's talents, as well as between the embodied experiences of being in the natural environment and the disembodied experience of sitting in front of a computer screen with its often violent simulation games that deaden the capacity for empathy and moral responsibility.

There are two other characteristics of the cultural commons that have special significance. The first is that they exist in every community and can be fully recognized only by an in-depth description of the cultural patterns that unconsciously influence the experience of preparing and sharing a meal, playing a game, telling a story, writing poetry, marching in an anti-war demonstration, protesting experimentations and other forms of animal exploitation, working with others in renewing habitats, and so forth. The cultural commons are largely taken for granted and thus unrecognized aspects of daily life—and can best be brought to attention through actual participation and ethnographic/phenomenological descriptions rather than through abstract theory and print-based descriptions. The second characteristic that needs to be reiterated, especially in light of the rate of global warming, is that what the industrial culture had to destroy, as Sale put it, are the intergenerational traditions that have smaller adverse impacts on the ecological systems.

Most aspects of the cultural commons in western countries rely to some degree on what has to be purchased. However, even this small degree of dependence makes a great deal of difference in terms of meeting the criteria of eco-justice. By being more intergenerationally connected, a revitalized cultural commons *reduces* the need for a system of production that has to dispose of vast amounts of toxic wastes (usually in the neighborhoods of the poor and marginalized). It also reduces the need to exploit the resources of Third World cultures and to integrate them into a global market system. As these cultures are able to regenerate their own cultural commons they are able to resist more effectively the West's efforts to colonize them in the name of development, democracy, and modernization—god-words that are based on western assumptions about individualism, progress, and the messianic drive to impose a consumer-dependent lifestyle on other cultures. The lifestyle that is more oriented toward cultural commons skills and activities of mutual support, and less on consumerism that degrades the environment and thus the prospects of future generations, meets yet another concern of eco-justice advocates. In possessing the skills and participating in the community systems of mutual support, the individual is more likely to resist the market-oriented ideology that equates the exploitation of species and habitats with progress. This characteristic of the cultural commons meets the last criteria of recognizing that natural systems have a right to reproduce themselves as part of the layered nesting of interdependent ecosystems—and not to be reduced to an economic resource.

This list of the ecologically sustainable and morally coherent characteristics of the cultural commons brings out what is missing in most of the educational discourse on how to eliminate discrimination in the areas of class, race, and gender. It also brings into focus the viable alternatives for addressing the estimated one billion lives that exist on one dollar a day, and are mired in the culture of poverty marked by a lack of food security and adequate housing. As global warming accelerates in the next few decades, as the world's oceans become less reliable sources of protein, and as droughts and severe weather systems contribute to mass migrations of people, the lives of the poor will become even more desperate as they expand in number. The double bind of relying upon sources of energy to keep the industrial system expanding (thus accelerating the rate of global warming) will intensify the willingness of corporations to outsource production facilities not only to low-wage regions but also to regions that still have easily accessed sources of energy. As the ecological crisis deepens, and the seemingly unrelenting drive to continue expanding profits in an increasingly stressed world becomes more difficult, it will be the people who continue to occupy the bottom rung of the economic/political/educational hierarchy who will continue to suffer the most.

The irony is that the ancient pathway of human development that still exists in rural and urban communities, and that represents an essential part of a post-industrial alternative, continues to be ignored—even by the few educational theorists who are beginning to recognize the ecological crisis. What now has to be avoided is the endless repetition that there is an eco-logical crisis and that capitalism is primarily responsible. Thoughtful people already understand the connections between these two phenomena. Instead, advocates of social justice need to explore the pedagogical and curricular implications of how to introduce students, including the already marginal-ized students, to the life-enhancing possibilities that exist in the cultural com-mons of their local communities—and that are part of the cultural commons of the dominant culture that protects the rights of various minority cultures. There is a direct connection between the enclosure of the traditions of civil liberties that are the basis of democracy and the growing dominance of cor-porations, market liberal politicians, religious fundamentalists, and the mili-tary establishment that views its mission as protecting the global interests of market liberals. There is also a connection in America between the number of marginalized groups who suffer the most deaths and catastrophic injuries from military actions that result from the logic of economic globalization. Knowledge of how to protest against the various forms of economic and political oppression is also part of the cultural commons—which includes the narratives of past protest movements, strategies that have proved most successful, and even the protest songs and iconography associated with past peace movements. The current drive to install total surveillance systems of a country's citizens as a defense against terrorism further undermines their civil liberties, and will be used by police to identify the leaders of movements pro-testing various forms of social injustice—including environmental activists.

Pedagogical and Curricular Implications

The future prospects of the poor and marginalized are inextricably tied to the future prospects of the cultural and environmental commons. With the outsourcing of work, computer-driven automation that reduces the need for workers, and downsizing in order to improve corporate profits, the prospects of upward mobility that have been an expectation of past generations, though unevenly realized, are being rapidly diminished. Given this reality, placing greater emphasis on educational reforms that help to regenerate the cultural commons should not be interpreted as meaning that all students, regardless of social class and ethnic background, should not acquire the knowledge that will enable them to find meaningful work that supports a basic standard of living. Just as most aspects of the cultural commons require some degree of dependence upon the industrial system of production and consumption, public schools and universities need to ensure that the students at the bottom of economic and social pyramid have the opportunity to learn what is required for careers and employment that are non-exploitive. At the same time, changes need to be introduced at all levels of the educational system that will enable students to learn about the community-centered alternatives that contribute to the transition to a post-industrial future—namely, the cultural commons. In discussing the unique characteristics of a pedagogy and curriculum that introduce students to the ecological and community-sustaining importance of the cultural commons, it is important to keep in mind that we are in a transition phase of cultural development. Thus, the following discussion of pedagogical and curriculum reforms must also be viewed in this light.

If we consider the basic tension between the industrial/consumer-oriented culture and the characteristics of the cultural commons that strengthen mutual support, develop skills and personal talents, and ensure moral reciprocity among all members of the community, it becomes clear what the role of the classroom teacher/professor should be. Instead of promoting the high-status forms of knowledge and values that contribute to the further expansion of the industrial/consumer-oriented culture, the role of the classroom teacher and university professor should be that of a mediator who helps students become aware of the fundamental differences between participation in the cultural commons and the culture of industrial production and consumption. Being a mediator requires an understanding of what students are most likely to take for granted as they move daily between participation in the two sub-cultures. The pedagogical task is to encourage students to name what would otherwise be taken for granted. Naming taken-for-granted patterns of thinking and behavior, as we learned from both the feminist and civil rights movements, is the first step to making them explicit, which is essential for developing communicative competence. Like the mediator in labor disputes, the teacher's mediator role precludes giving students the answers about which aspects of the cultural commons and the industrial/consumer-oriented culture need to be rejected or renewed. The techno-scientific basis of the industrial culture has made many important contributions to improving the quality of human life, and now has the potential to help reduce our carbon footprint. Thus, the task of being a mediator should not be reduced to that of an ideologue who has pre-conceived answers, and who enforces the silence about what her/his ideology cannot explain. Similarly, ideology should not guide how the students are to think about their embodied experiences within the cultural and environmental commons.

The initial step in teaching and learning that fits the model of a mediator is to encourage students to describe their embodied/culturally-influenced experiences as they move between the two sub-cultures. There are specific questions that students need to be reminded to ask: Does the experience in a cultural commons activity contribute to the development of personal skills and the discovery of talents? Does it contribute to a sense of community self-sufficiency and mutual support? Does it require exploiting others who are less advantaged? What is its impact on natural systems? Does it contribute to an awareness of what needs to be intergenerationally renewed and of the need to be able to mentor others? Does it lead to different forms of empowerment, such as the ability to exercise communicative competence in resisting further forms of enclosure of skills and patterns of mutual support that result in an increased dependency upon a money economy? What is its ecological footprint? These same questions need to be explored by students as they participate in various aspects of the industrial/consumer-oriented culture.

In examining the differences in experience between preparing and sharing a meal with others and eating in a fast food outlet; between face-to-face communication and reading; between gardening and being dependent upon industrially prepared food; between participating in one of the creative arts and being a consumer of commercially promoted artistic performances; between developing skills associated with a craft that extends one's talents and purchasing what has been industrially produced (increasingly in a low-wage region of the world); the differences in personal development will quickly become apparent. And this awareness of differences, if framed in light of the ecological crisis and the changes resulting from economic globalization, is essential to the recovery of local democracy that has been one of the hallmarks of the diverse cultural commons that have not been based on ideologies and mythologies that have privileged the few over the many.

Another responsibility of the teacher/professor's mediating role is to ensure that students become aware of the narratives that provide an account of various social-justice movements—starting with the earliest beginnings of the traditions of civil liberties in the West. These include habeas corpus, the right to a fair trial by a jury of peers, separation of powers, and an independent judiciary. The narratives that provide an understanding of the labor movements that struggled to achieve safe working conditions, a living wage, and the right of workers to organize politically, should also be part of the curriculum. The feminist as well as the civil rights movements also should be part of a commons-oriented curriculum. Again the tensions between the cultural commons and the industrial/consumer-oriented culture that are now being globalized, and that are major contributors to the ecological crisis, will inevitably come out—and be a major focus of class discussions.

The ecological crisis, as well as the increasing number of the world's population that is moving from a subsistence existence into one of dire poverty, make it particularly important that the teacher/professor introduce students to the history of different ways in which the cultural commons are being enclosed. The following questions will bring into focus different forms of enclosure: How did the western philosophers' reliance on unacknowledged culturally-influenced interpretative frameworks (which can also be understood as root metaphors that frame the historically layered process of analogic thinking) contribute to the enclosure of the cultural commons? How has the rise of western science contributed to the enclosure of local knowledge of healing, agricultural practices, reliance on local materials, and so forth? What role have various religions played in strengthening the cultural commons and, on the other hand, in representing the exploitation of the commons by market forces as carrying out God's plan for those who are to be saved? What were the intellectual influences that marginalized the importance of the workers' skills, their control of the tempo of work and use of technologies? What are the current techno-scientific and market forces that are threatening the genetic diversity of seeds, and local knowledge of how to adapt agricultural practices to the characteristics of local soils and weather patterns?

In addition to introducing, particularly as the students move into the upper grades and onto the university, the various histories of different forms of enclosure, the role of being a mediator also requires that students be introduced to how different cultures have sustained their cultural and environmental commons while at the same time ensuring that their local markets did not dominate the patterns and values of everyday life (W. Sachs, 1992; W. Sachs, 1993). Knowledge of the intergenerational traditions of other cultural approaches to the cultural and environmental commons will enable students to gain a better perspective on whether the current myth that equates the western scientific-technological and market-driven approaches to creating greater dependence on what is industrially produced and consumed should be the basis of development in other cultures. There is a need to enable a large percentage of the world's population that is mired in poverty to obtain a decent standard of living and to enable them to experience more than a life of drudgery and stunted development. The critical question is whether the further enclosure of the diversity of the world's cultural commons will achieve this end.

To this point, the discussion of the teacher/professor's role as a mediator between the students' culturally embedded experiences in the local cultural commons and in the workplace and shopping malls of the industrial culture has been general in nature. It is now necessary to address how to engage students from a variety of backgrounds that make them especially vulnerable to the prejudices currently perpetuated by the educational system's emphasis on the high-status knowledge that perpetuates poverty and deepens the ecological crisis. As mentioned earlier, every culture has its own intergenerational traditions of preferred foods, approaches to the creative arts, healing practices, ways of understanding moral reciprocity, craft knowledge, narratives of past achievements and leaders, mentors in various arts and crafts, understanding of social justice, and so forth. For example, in the largely Hispanic community in San Francisco one will find that many of the walls of buildings that previously were used to advertise cigarettes and liquor have been reclaimed as part of the cultural commons. Giant murals now depict past struggles, important cultural leaders, and visions of what the future should hold for Hispanic communities.

The same reclaiming of this part of the cultural commons can be found in Detroit and other major cities. Other examples of the cultural commons can be seen in the community gardens where traditional foods are grown, in the local poets, artists, writers, and musicians who are willing mentors of the community's youth. There are elders and people who take responsibility for keeping alive the oral history of the group, just as there are living traditions of how to assist the especially vulnerable to the problems of extreme poverty, old age, and hopelessness. The nature of these cultural commons varies from community to community, from ethnic group to ethnic group. As the cultural commons of these ethnic and marginalized groups are nested in the cultural commons of the larger society, with its traditions of civil liberties, of achieving legal redress of discriminatory practices, and of effecting changes through an admittedly flawed democratic process, it is important that these traditions also be recognized as essential aspects of what marginalized students should claim as their cultural commons.

The starting point in a commons-oriented curriculum is to have students conduct a survey of their local cultural commons, as well as the aspects of the larger cultural commons that they have a right (in spite of past exclusions) to participate in. The survey should involve learning who the elders and mentors are, who the keepers of the community memory are, what forms of cultural commons activities exist—such as playing chess, painting, writing poetry, musical performances, gardening, working with wood and metal, volunteerism, and political action groups. In a word, the survey should cover the activities and relationships within the community that are less reliant upon a money economy—and that lead to the development of skills and interests that contribute to a less damaging ecological footprint.

After the survey has been undertaken, the process of learning to make explicit the differences between their culturally nested experiences with different activities within the cultural commons and in the world of industrial work and consumerism can begin. This process of learning to recognize differences that otherwise are taken for granted as the students move between the two sub-cultures, and to name them, provides the linguistic and conceptual basis for the communicative competence necessary for resisting further forms of enclosure by market and scientific/technological forces. Resistance may take the form of overcoming the silences about the nature and importance of the local cultural commons being perpetuated in public schools and universities. It also may take the form of resisting the false promises of developers who want to attract the large commercial enterprises that will eliminate the small shop keepers and service providers, as well as the open physical spaces that enable members of the community to connect with the natural world, to have community gardens and places for children and others to play, and to escape the pressures of the media and the temptations of the shopping malls. Communicative competence is also necessary to giving voice to what aspects of the techno-scientific/industrial culture need to be abandoned as ecologically unsustainable—and which aspects can make a contribution to improving the lives of people while still having a smaller ecological footprint.

One of the failures of the educational theorists who have been writing about the need for educational reforms that address the seemingly intractable problems of class, race, and gender discrimination is that they have continued to use the metaphors of *individualism, progress, emancipation, intelligence, tradition,* and so forth, that carry forward the analogs formed in the distant past by theorists who ignored cultural differences, the nature and importance of the cultural and environmental commons, and the existence of ecological limits. In effect, the arguments for addressing the issues of race, class, and gender have been based on a metaphorical language that has been frozen over time, and that continues to put out of focus the intergenerational relationships and knowledge that provide alternatives to the form of individualism that is dependent upon consumerism to meet daily needs. Reliance upon the metaphorical language that gave conceptual legitimacy to the rise and current globalization of an industrial/consumer dependent lifestyle can also be understood as yet another unrecognized example of how language continues to colonize the present by past ways of understanding.

Learning to participate in what remains of the local cultural commons, and in developing new skills and non–monetized relationships will have the effect of expanding how intelligence is understood—from that of an individual attribute that is subjectively centered to understanding that intelligence is communal, intergenerational, and enhanced through participation with others, and with the environment. As the communal and intergenerational nature of intelligence may be the source of prejudices and environmentally destructive lifestyles, it is important that teachers/professors help students recognize the forms of intelligence that are destructive of human possibilities, as well as the ways of thinking that are informed by today's understanding of social and ecojustice. This has been part of the curriculum that addresses various forms of discriminatory relationships and patterns of thinking. Too often this emphasis on emancipation has reinforced the idea that critical thinking is the expression of individual autonomy. Making the students' culturally-influenced experiences in the local cultural and environmental commons an integral part of the curriculum will help reconstitute how individualism should be understood—from that of being autonomous and essentially alone to recognizing that one of the unique characteristics of life is being in an ecology of relationships that constantly lead to a redefinition of self that reflects changes in the social and environmental context. The word "tradition," which still carries forward the reductionist thinking of Enlightenment writers, will also cease to be an abstraction that misrepresents the complexity of daily experience in both the cultural commons and in the industrial/consumer-oriented culture. Instead of thinking that change is always a progressive force, the students' reflections on their experiences within both sub-cultures will lead to a more complex and critically informed understanding of which traditions need to be carried forward and renewed, and which traditions need to be rejected as environmentally destructive and as sources of injustice.

One of the metaphors that is in special need of being associated with new analogs is *environment,* which is now understood either as the background within which human experience takes place or as an exploitable resource. If the teacher/professor explains, and has students test out in terms of their own embodied experiences, how different environments can be understood as ecologies—and that ecologies include both the interactions and interdependencies within natural systems as well as within cultures (and the interdependencies between culture and nature) students are more likely to be aware of the different ways in which their activities impact the sustainable characteristics of natural systems. Students still rooted in the beliefs of their indigenous heritage already possess this awareness, but students who have been uprooted from their cultural traditions, which may not have been ecologically-centered in the first place, will need to develop this awareness. And this awareness will be essential to slowing the rate of environmental degradation that will impact them the hardest in coming years.

The challenge now is for the proponents of educational reforms that address the issues of class, race, and gender to recognize that an approach to achieving social justice for the millions of marginalized students cannot be based on the same deep cultural assumptions that created the industrial/consumer-oriented culture that is largely responsible for the injustices that continue to stunt the potential of students. This challenge will be particularly difficult to address as few of today's proponents of educational reform have given attention to how language helps to organize their own patterns of thinking in ways that reproduce the silences and cultural assumptions of past theorists who contributed to today's double bind patterns of thinking. The problem is that the double bind thinking of these self-proclaimed social justice theorists continues to equate progress with achieving greater equality of opportunity for marginalized groups to live a middle class consumer dependent lifestyle—while the world is moving closer to the ecological tipping point scientists are warning about.

References

Ayers, William, Theresa Quine, and David Stovall (editors). (2009) *Handbook of Social Justice in Education.* New York: Routledge.

Bowers, C. A. (2011). *Educational Reforms for the 21st Century: How to Introduce Ecologically Sustainable Reforms in Teacher Education and Curriculum Studies.* Eugene, OR. The Eco-Justice Press.

McLaren, P. 2005. *Capitalists and Conquerors: A Critical Pedagogy Against Empire.* Lanham, MD: Rowman & Littlefield Publishers.

Moyers, B. (2007) Bill Moyer's journal, August 3, 2007. Available at www.pbs.org/moyers/journal/08032007/transcriptsl.html.

Polanyi, Karl. (2001 edition) *The Great Transformation.* Boston: Beacon Press.

Sachs, Wolfgang. (editor) (1992) *The Development Dictionary: A Guide to Knowledge as Power.* London: Zed Books.

_____. (editor, 1993) *Global Ecology: The New Arena of Political Conflict.* London: Zed Books.

Sale, Kirkpatrick. (1995) *Rebels Against The Future: The Luddites and Their War on the Industrial Revolution.* Reading, MA.: Addison-Wesley.

Shiva, Vandana (1996) *Protecting Our Biological and Intellectual Heritage in the Age of Biopiracy.* New Delhi: Research Foundation for Science, Technology and Natural Resource Policy.

Part IV
Critique of Technology

13 Educational Reforms that Contribute to Democratizing the Uses of Digital Technologies

There is nothing in the diverse realms of communication that is politically neutral. The earlier discussion of the differences between print-based cultural storage and communication and oral traditions brought out the fact that there are gains and losses at all levels of cultural life. Similarly, the use of English nouns that reinforce the illusion of fixed entities and ideas cannot, at the same time, reproduce the dynamic relation/information-rich processes that are essential characteristics of cultural and natural ecologies. These fixed representations reduce our capacity to exercise ecological intelligence—which involves being aware of what is being communicated through relationships. The metaphors that carry forward the misconceptions and silences of earlier eras become the basis of the linguistic colonization of the present by the past. The amplification and reduction characteristics of technologies also point to the inherently political nature of digital technologies—whether they take the form of robots on an assembly line, software programs that enable security agencies to track the behavior of a massive number of cell phone users, or that collect data and model the connections between the warming of the oceans and the movement of fish populations that will reduce the protein available for millions of people.

That all forms of communication have political implications should be part of the common-sense knowledge of everyone, and not just what a few academics write about. The political nature of digital technologies also should be widely recognized by even the most casual observer. Indeed, it would seem that the titles of books by computer scientists and futurist thinkers, such as Kurzweil's *The Spiritual Machine*, and *How to Create a Mind*, Eric Schmidt and Jared Cohen's *The New Digital Age: Reshaping the Future of People, Nations, and Business*, and the subtitle of Gregory Stock's *Metaman: The Merging of Humans and Machines into a Global Superorganism*, should have set off alarm bells throughout the halls of universities where the departments of journalism, political science, sociology, anthropology, and philosophy are located. The strong imperialistic message in these titles should have led to widespread questioning in the media and within our educational institutions about whether the computer scientists and the corporations that market every new technology have consigned democracy to the graveyard of outmoded cultural "memes" (to use one of their neo-Darwinian metaphors).

Part of the task here is to understand why computer scientists and their futurist thinking supporters are given near total freedom to change the world in ways that serve the interests of corporations, and their own narrowly conceived imperialistic interest in making more of daily life dependent upon their technologies. Granted, there are many important and now indispensable uses of digital technologies—and the convenience they provide for the average citizen is a major reason why so few people are asking where this emphasis on technology is leading us. But there are other reasons that digital technologies are embraced so widely and enthusiastically that go beyond the convenience and problem-solving ability they afford. The two that will be addressed here are the failures of our educational institutions and the continued domination of corporate capitalism. As many have already written on the social injustice issues and, now, the ecologically destructive nature of corporate capitalism, I shall quote Jerry Mander's (2012) summary of the limitations that should already be widely recognized—except for the power of the media and the mythic foundations of modern thought. I shall then turn attention to the educational reform issues that need to be addressed.

In *The Capitalism Papers: Fatal Flaws in an Obsolete System* (2012), Mander frames the arguments against corporate capitalism in the following ways:

- Amorality. Capitalism's only purpose and mandate is the expansion of individual and corporate wealth. It has no other job, and no interest in "right or wrong," or human welfare, or communities, or in the well-being of the natural world, except as resources for itself.
- Dependence upon growth. This is the most fundamental problem, though it is the least noticed, and is rarely included in mainstream economic and political discourse. The entire system and its practitioners, like all of modern society itself, have become seriously oblivious to any connection with nature and its limits, or to the realization that human activities are embedded in a larger natural system that has been under deadly assault for centuries. No solutions to the crises today will be possible if the system ignores this fundamental reality, or is unable to correct it.
- Propensity to war. This includes buildup to wars, innovations for wars, rebuilding after wars, and forward basing for defense against future wars. It encourages what some now call "permanent war" and is viewed as a highly effective economic strategy, good for both wealth creation and jobs. Where others see destruction, capitalism sees opportunity.
- Intrinsically inequitable. In every fiber of its structure, from its strictly hierarchical structural forms to its practical performance, the system expresses and expands inequality. The central function of capitalism is to help people with wealth to seek more wealth and greater dominance; the

separation between rich and non-rich within countries and among them inevitably become steadily greater. We have achieved plutocracy. The Occupy movement's people are on the case.

• Undermines democracy. The system has an intrinsic need to dominate and undermine democracy, and also public consciousness, so as to control its rules, benefit more easily, and advance its primary self-interest: expanding growth, profit, and wealth. Governments become subordinate, and democracy is destroyed.

• Capitalism does not bring happiness. Societies based on a constant quest for external satisfactions, such as wealth advancement, commodity accumulation, and competitive advantage—quests for "prosperity and power"—do not tend to achieve overall well-being; quite the opposite. Happiness and well-being are rooted in other values and behaviors.

(Mander, 2012, p. 13)

It needs to be emphasized that his criticisms are aimed at corporate rather than at what he refers to as hybrid or community-centered capitalism where small-scale economic activity is guided by values that strengthen communities and that have a smaller ecological footprint.

The Continuing Hold of Archaic Ways of Thinking that Education Reforms Must Address

The lack of education about the cultural non-neutrality of technology, specifically digital technologies, is one of the reasons why most of the public and even most academics across the disciplines remain silent in the face of technological imperialism. This technological imperialism envisions a world in which machines eliminate the need for workers and in which the diversity of cultural knowledge systems are to be replaced by super-intelligent machines. The linear direction of cultural transformation they read into the Rosetta Stone of what has been turned into neo-Social Darwinism aligns with the western assumptions about the connections between change, especially technologically driven change, and the linear expression of progress.

Two points need to be made about the possible reasons why so few questions are being raised about the cultural transforming influences of digital technologies. As pointed out in an earlier chapter, the deep taken-for-granted cultural assumptions about individualism, a human-centered world, the cultural neutrality of technology, the need to understand traditions as limitations on further progress and thus in need of being overturned, the importance of monetizing relationships, achievements and patterns of social interaction, and so forth, are largely shared by both the mainstream in American society and by computer scientists. The other possible explanation for the silences in the face of this world-transforming technology, beyond the fact that it has many important uses—which leads to ignoring its destructive influences—is that most current academics, including computer scientists, were heavily influenced by the silences in the education of their mentors whose thinking was framed by the orthodoxies of the late 20th century—and earlier.

The most critical characteristics of the silences about the political nature of language that were largely ignored in the last decades of the 20th century can be traced to the literature that would have radically challenged the myth that there is such a thing as objective knowledge. This literature was just beginning to appear in the late 1970s and 80s. The widely held idea that print is able to represent objective knowledge was supported by thinking of language as a sender/receiver process of communication. That is, language was understood as a conduit through which ideas, data, and information could be sent to the reader. Michael Reddy's classic article, "Conduit Metaphor: A Case of Frame Conflict in Our Language about Language" (1979) represented a fundamental challenge to this widely held orthodoxy. Reddy's challenge of one of the most important cornerstones upholding the edifice of objective knowledge came at a time when words were not recognized as metaphors that encoded earlier cultural assumptions that influenced the choice of analogs used to frame the meaning of words.

Reddy opened minds to the possibility of radically rethinking the nature of words that supposedly represented the objective knowledge being sent through the conduit of language. Among the most important books that broke the spell was George Lakoff and Mark Johnson's *Metaphors We Live By* (1980). The still not fully digested classic of Gregory Bateson, *Steps to an Ecology of Mind* (1972), which contained his statement that the "map is not the territory," had preceded Reddy's groundbreaking essay. To stay with Bateson's metaphor, the "territory" is the current condition of the cultural and natural ecologies within which we live. This was Bateson's shorthand way of pointing out that most words are metaphors and that their meanings are framed by analogs settled upon in the distant past, and thus are unreliable for thinking about the state of current ecological systems. In effect, Bateson's insights about the recursive patterns of thinking brought into focus the fact that words (metaphors) have a history and thus carry forward earlier misconceptions, silences, and culturally specific assumptions. This insight is ignored by nearly all classroom teachers and most university professors. And it will continue to be ignored as online education becomes more widely adopted.

While Lakoff and Johnson (1980) made the case in *Metaphors We Live By* that all thinking is based on interpretative frameworks that are themselves metaphorical, they did not understand the nature of root metaphors, including how root metaphors such as patriarchy, individualism, progress, mechanism, evolution, ecology, and so on, exclude vocabularies essential to challenging the cultural-shaping root metaphors. Later, in their jointly authored book, *Philosophy in the Flesh: The Embodied Mind and Its Challenge to Western Thought* (1999), the allure of aligning their theory of metaphor with the field of cognitive science led to undermining the awareness that words (metaphors) have a history, as well as their role in the linguistic colonization of other cultures. Two of the six guidelines they identify as essential for understanding the origin of metaphors included the following: "Embodied Concepts: Our conceptual system is grounded in, and neurally makes use of, and is critically shaped by our perceptual and motor systems." Secondly, "Conceptualization Only Through the Body: We can only form concepts through the body. Therefore, every understanding that we can have of the world, ourselves, and others can only be formed in terms of concepts shaped by our bodies" (Lakoff and Johnson, 1999, p. 55).

What they termed "embodied reason" highlights the fact that many of our metaphors are indeed derived from bodily experiences, but they abandoned what is most needed to understand the historical influences on today's double-bind thinking, where we repeat the mistake Einstein warned about: namely, relying upon the same mindset (language system) that created the crisis to fix it. If we are to understand the connections between language and the ecological/cultural crises, including how the digital revolution perpetuates the conduit view of language that hides its colonizing role, we will have to begin taking Bateson seriously.

The dependence upon the cultural commons, that is, the intergenerational knowledge and skills that have been shared on a largely non-monetized basis since the first humans were walking around the savannas of what is now referred to as Africa, was not recognized in Roman Law and in the Magna Carta of 1215. While daily life was dependent upon the traditions of the cultural commons, only the environmental commons were recognized. The idea of the commons popularized within environmental circles by Garrett Hardin's essay "The Tragedy of the Commons" (1968) led to an equally restricted and ethnocentric view of the commons. Paul Goodman's *Growing Up Absurd* (1961) and Ivan Illich's *Deschooling Society* (1971) and *Tools of Conviviality* (1973) challenged the idea of public education on the grounds that it excluded the practical knowledge needed in self-reliant communities that represented alternatives to a consumer-dependent lifestyle, but they did not introduce the idea of the cultural commons or the concept of enclosure.

As I argued in a number of books (*Let Them Eat Data*, 2000; *University Reform in an Era of Global Warming*, 2011; and *The Way Forward*, 2012), both concepts are essential to understanding that the non-monetized intergenerational knowledge of the cultural commons, as well as the technological and ideological forces of enclosure, should be one of the primary foci of educational reform. Indeed, the ways in which western technologies, when driven by the market liberal ideology, contribute to the enclosure of the cultural commons deserves special attention. We have long understood how industrial technologies undermine craft knowledge and skills, but we are just beginning to understand how digital technologies are eliminating the need for workers. What continues to be ignored is how the vast numbers of unemployed people are to sustain themselves economically, and to have a meaningful life. Critics of the idea of introducing cloud computing and

3-D printing into the garment industries in Bangladesh and other low-wage regions of the world estimate that 50,000 garment workers (mostly women) would lose their source of income. As social justice issues have not influenced a wide range of current efforts to displace humans with machines, it is unlikely that issues of poverty will trump the possibility of being able to earn more profits by modernizing the production process. This is where the importance of revitalizing the local cultural commons and learning to resist the various forms of enclosure that are driven by market liberal ideologies, technologies, and the silences on the part of educational elites needs to be understood.

During the 1970s and 80s when a radical rethinking of the ecology of language was just beginning, a number of books that examined the differences between print and oral-based cultures appeared. The most prominent included Jack Goody's *The Domestication of the Savage Mind* (1977), Walter Ong's *Orality and Literacy: The Technologizing of the Word* (1982), Eric Havelock's *The Muse Learns to Write: Reflections on Orality and Literacy from Antiquity to the Present* (1986), and Ivan Ilich and Berry Sanders' *ABC: The Alphabetization of the Popular Mind* (1988).

The old saying about cultural lag applies to how the taken-for-granted patterns of thinking that were represented as high-status, cutting-edge thinking in the last decade of the 20th century continue to be the basis of thinking among most of today's environmentalists, computer scientists, market liberals, and politicians who are focused on shaping the future. As pointed out in earlier chapters, the computer scientists and promoters of the digital takeover of the world's cultures were not influenced by the fundamental shifts in thinking introduced by Bateson, Reddy, Ong, and the others. Thus, they continue to ignore how they remain trapped in the old epistemological/linguistic patterns of earlier centuries that continue to give conceptual direction and moral legitimacy to the digital phase of the industrial/consumer-dependent culture that still does not take account of environmental limits.

Unfortunately, few academics across the disciplines were influenced by the new potentially cultural-altering insights of the 1970s and 80s, with the result that their students, now occupying key positions in shaping public thinking, continue to ignore the cultural roots of the ecological crisis, continue to think of language as a conduit and to ignore that words are metaphors and thus have a history. They also continue to ignore the culturally diverse evidence that technologies, especially print and now electronic-based technologies, are not culturally neutral. Unfortunately, we are now so far down the road of being dependent upon digital technologies that the recent books by Nicholas Carr, Robert W. McChesney, and Sherry Turkle that question the cultural transforming nature of digital technologies may be too late to reverse the colonizing and ecological changes now underway.

Also unknown to most graduates of higher education is the vast literature on the cultural transforming nature of western technology. The major thinkers whose influence has been largely negated by the modernizing orthodoxies and silences that have served as the basis of the high-status knowledge promoted in universities that supported the expansion of the industrial/consumer-dependent culture include the following: Karl Marx, Martin Heidegger, Paul Ricoeur, Jacques Ellul, Lewis Mumford, Ivan Illich, Peter Berger, Theodore Roszak, Michel Foucault, Don Idhe, and Langdon Winner, among others. Granted, the technology/cultural issues raised by these authors would make for heavy reading, even in past decades when the attention span of students could tolerate extended explanations and analyses.

The challenge today is to introduce key ideas and issues in ways that will engage the attention of the wider public, who must be brought into the process of democratizing the introduction of new technologies. Unlike the older generation whose attention span was not shortened by the consciousness-shaping influence of the Internet and by the constant need to check their cell phone or Twitter for an incoming message, but were never challenged to question the myth that held that technology is culturally neutral, the cyberspace generation will need to be engaged in a profoundly different way. If the myths surrounding technology are to be seen for what they are, engaging the cyberspace generation must avoid the requirement of reading long and seemingly dry tomes that often represent the technology/culture nexus in overly abstract ways. What is being suggested here will require providing short explanations of basic processes and relationships that are missing in most discussions of technology, followed by the identification of key insights, issues, and problems that students will be asked to consider in terms of their own culturally influenced experiences.

What is distinctive and potentially workable about this approach is having students give voice to their own insights about the relationships they experience as participants in the technology-influenced cultural ecologies. To avoid the pitfalls that come with emphasizing that students should articulate their own insights, and stake out what they want to learn, which too often is uninformed by an understanding of the larger cultural/ecological contexts, it may be necessary to continually restate the important issues as part of the larger ecology of thinking—including what they may not recognize because of the silences in their own education. That is, the classroom teacher, university professor, and discussion leader need to be able to name the cultural patterns and relationships of which students may not be aware because of their taken-for-granted status. This can be done without imposing an ideologically driven interpretation as the process of naming the patterns and how they interact with each other should stop short of explaining what the student should be experiencing. If everyone shares the same taken-for-granted cultural patterns of behavior and thinking, the discussion will touch on only what everybody already knows—and this will soon be seen as a waste of time for everyone.

One of the issues I have encountered that led to the above suggestions is that in introducing ideas related to how current cultural/linguistic processes are ecologically problematic, the Other (the individual or the audience) initially appears to indicate both understanding and concern. Unfortunately, there is seldom any carry-over in terms of further reflection and action after the presentation. It is too often experienced as a short performance, like watching a television program, where viewers move on to other daily concerns after turning off the program. The reason, as I have observed how audiences enter into what becomes for them a conceptual and moral cul-de-sac, is that they lack the larger conceptual frameworks that would enable them to take the next steps in ecologically sustainable thinking and behavior on their own.

The long-held individually centered pattern of thinking involves the assumption that events, ideas, and issues are discrete and can be understood on their own terms; whereas an ecologically informed pattern of thinking requires recognizing that everything, ranging from ideas, behaviors, issues, and so forth, can be more fully understood within the larger ecology of interdependent cultural patterns—including ways of thinking. For example, to understand what is problematic about print, one needs to understand the emergent/relational nature of cultural and natural ecologies. To understand what is problematic about individual intelligence, it is necessary to understand the different levels at which we exercise ecological intelligence. And to understand what is problematic about the idea of progress or the implications of robots displacing the need for workers, it is necessary to consider the many ways in which the western form of progress are overshooting the capacity of natural systems to renew themselves, and to ask how the massive number of unemployed people will be able to support themselves.

Most people, including many academics, do not understand the nature and ecological importance of revitalizing the world's diversity of cultural commons, including their languages, because they continue to think of the needs and behaviors of individuals within the context of a consumer-dependent economy. And they do not understand what is problematic about print-based storage and communication because they ignore the complex nature of oral communication, and so forth. Presenting new ecologically informed ways of thinking too often encounters the conceptual cul-da-sac mentality—and a lack of curiosity as to how to break out of it. When more areas of the earth are being overwhelmed by droughts, floods, and other extreme weather conditions that limit access to clean water, to sources of protein, and that limit the ability to meet other basic needs of life, the larger public may wake up to an awareness that they should learn to think ecologically—but then it might be too late.

Two Key Overarching Conceptual Frameworks: The Nature of Ecological Intelligence and the Linguistic Construction of Reality

The following is an attempt to adapt to the demands of the emerging digital-conditioned mindset that has a short attention span and thus can seemingly cope only with short explanations—which is itself a threat to democracy. The concepts introduced here are intended to redirect students' attention to the cultural processes and patterns that are part of their everyday experience. Giving attention to these background cultural patterns and giving voice to them, in turn, will provide the conceptual basis for understanding the colonizing and thus cultural non-neutrality of digital technologies.

1) *All Living Systems are Ecologies—Including the Exercise of Intelligence*
(A Review of Core Understandings that should Precede the Following Questions and Discussions)

Contrary to a dominant assumption in western thinking, there are no isolated entities, organisms, individuals, or ideas. Everything exists in relationships, and understanding them—from the behavior of a cell to that of an individual, and to that of macro ecological systems such as the behavior of oceans—requires understanding what is communicated through the information pathways (relationships) and how it elicits a response that, in turn, leads to further interactive responses. This understanding should serve as the background conceptual framework for all subsequent discussions of the non-neutrality of technology.

Students in the earliest grades can be asked the following questions that will enable them to experientially ground the basic concept that nothing exists in isolation, and that understanding the Other (a person, plant, word, event, idea, and so forth) should take into account what is communicated within and between the other participants that make up the ecological system. It is important that they understand that the form of information exchange will vary between organisms—for some it will be at the chemical level, others will respond to temperature differences, while for animals and humans it may include all of the above and also include non-verbal patterns as well as the spoken word and other culturally coded systems of signs. The point often overlooked because of the human-centered view of what constitutes communication is that everything in the cultural and natural environment communicates, with some of the communication being at the level that exceeds what humans can be aware of—such as the scent left by other animals, sounds and what they mean to other animals, and so forth. The starting place is to ask students to consider the following questions about the nature of living systems and whether they are part of them—or stand apart as an autonomous individual:

Question #1: Can you identify anything that is totally isolated—that has no relationship with anything else?

Question #2: What are the relationships that currently affect your experience? (Teachers should push them to go beyond identifying friends, sports, cell phones, parents, and so forth. Ask them about environmental influences, language, and other relationships they may not have considered. The question should lead to a deep ethnography—or thick description—of their immediate experiential situation).

Question #3: What is communicated through these relationships, and how does your response lead to further changes in what is being communicated through the relationships of which you are aware? Are the relationships limited to other people? What is communicated through your relationships with the natural environment? Can you describe examples of the interactive nature of your other relationships—in playing a game, in a conversation, in interacting with an animal, in planting a garden, in your use of water, in eating food flown in from another country? After identifying other living entities, ask what relationships affect life-sustaining processes—such as the relationships between toxic chemicals and the behavior of bees, between the use of different technologies such as genetically engineered seeds and the behavior of the farmer, between the behavior of the person driving a car and changes in the condition of the road and the behavior of other drivers, between changes in global temperatures and the melting of glaciers, etc. It is important to clarify that relationships involve not only interactions but also information coded in the differences which make a difference—and that these differences which make a difference involve multiple messages that affect life-sustaining and altering processes.

Following a discussion of whether students have themselves experienced contexts that are free of all relationships within the cultural and natural environment, they should be encouraged to discuss the cultural influences that reduce awareness of the complex message systems that are an inescapable part of living in cultural and natural ecological systems. Does abstract thinking, an emphasis on data, being culturally conditioned to think of oneself as an autonomous individual, relying upon a vocabulary that represents the world as fixed and thus unchanging, etc., contribute to being unaware of what Bateson referred to as the differences which make a difference, and what Michel Foucault referred to as an "action upon an action"? It is also important to discuss the cultural influences on how they respond, largely at a tacit and taken-for-granted level, to many of the messages being communicated through the different information pathways of the ecological system in which they are participants. It should be mentioned that the study of natural ecologies by scientists involves the same focus on the interactive nature of what is communicated through the information pathways that constitute different forms of relationships—which may range from the molecular and chemical to macro systems, such as how changes in the temperature of oceans leads to changes in weather patterns, to changes in habitats, to changes in food production, to changes in diets, to changes in health. If students understand the behavior of ecological systems as they interact and influence other systems, they will be better able to avoid the linear pattern of thinking that the above example suggests.

This introduction to the nature of ecological systems will be important for later discussions of how digital technologies provide important data and information on the behavior of natural ecologies, but misrepresent the nature of cultural ecologies—including the nature of ecological intelligence, which is exercised when giving attention to the information pathways in ongoing and emergent relationships. When more attention is given to what appears on the computer screen, or to what is being texted, or communicated through the cell phone, less attention is being given to the ongoing relationships within daily life. Thus, more of life is being lived in the world of abstract images and words, which may be useful in some circumstances but at the same time undermines the exercise of ecological intelligence. This latter point should be the focus of further student discussions.

2) *What People Assume to be Reality is Largely Linguistically Constructed—And Thus Will Differ from Culture to Culture* (A Review of Basic Understandings that should Precede Raising the Following Questions)

Contrary to another dominant way of thinking reinforced in the West, there are no autonomous individuals or thought processes. The individual is born into a language community, and this language is largely metaphorical—which means it has a history and carries forward the misconceptions and silences of earlier eras when there was no awareness among western thinkers about environmental limits. Learning to think is influenced by the vocabulary (metaphors) shared within the language community. As others rely upon the same linguistically constituted interpretative frameworks, what is named and what is hidden or misrepresented by the metaphors that reproduce earlier thinking and value frameworks largely becomes the individual's taken-for-granted reality. Critical thinking that questions the discrepancies between the inherited interpretations of reality and personal experiences, including social injustices and environmentally destructive behaviors, often leads to new ecologically and culturally informed metaphors and a different understanding of reality. Understanding the relationship between language and how it constructs what becomes the taken-for-granted view of reality needs to be introduced in the later grades—including at the university level if the conceptual/ecological implications are to be understood at a more complex level.

Question #1: Depending upon the maturity of students in the earlier grades, they can be asked to identify the vocabulary that supported a male-dominated way of thinking—and the taken-for-granted behaviors that were consistent with this vocabulary. Also, they can be asked to identify how the vocabulary that represented nature in early children's stories differs from the vocabulary of today's environmentalists. In later grades, students should be asked to identify the many metaphors that support the idea of progress, that represent a mechanistic view of reality, that support the idea of the autonomous individual, that support a human-centered view of the world.

Question #2: Ask students to consider whether the different cultures represented by the students in the class associate the same meanings with words (metaphors) such as "individualism," "education," "progress," "intelligence," "wealth," "development," and so forth. Have them build a list of other metaphors that support the culturally shared understanding of what these words mean. This list will vary between cultures, which can be the focus of further discussions.

Question #3: Ask the students (in upper grades and even college level) if these metaphors have a history, and if their meanings have changed over time. What cultural changes led to changes in the meaning of these metaphors? Does the introduction of western metaphors into the vocabulary of non-western cultures become a form of cultural colonization? What are some examples?

Question #4: What are the experiential differences between print and the spoken word? What aspects of an experience—a conversation, a game, a long-term interaction with some aspect of the natural environment—get omitted from a print-based account? What are the important uses of print—in your own experience, and in understanding social issues and processes? Would computers be as useful if they did not rely upon print?

Review what students should take from examining the connections between these language processes and how they think about reality that will enable them to understand which cultural orientations are reinforced by digital technologies: that most words are metaphors and thus have a culturally specific origin that fosters a particular view of reality, that language is not a conduit through which objective information and data are passed, that print is highly useful in certain situations but that it provides only a surface and static representation of a reality that is dynamic and emergent, as is the case with all living (that is, ecological) systems.

The combination of questions and student explorations of what the questions lead them to consider about how various language processes influence awareness, including what they ignore, provides the conceptual basis for recognizing what is being ignored by computer scientists and programmers who assume that the digital culture represents progress for the entire world. These explorations of the language/culture/experience connections highlight the most basic relationships that should be considered in determining the appropriate and inappropriate uses of digital technologies.

3)   *Technology Issues to be Discussed with Older Students and Within Adult Learning Groups*

Again, the questions should lead to examining the lived cultural patterns that are otherwise ignored by computer scientists and programmers because of their taken-for-granted status.

Question #1: What are the differences between machine and human intelligence? How do differences in cultural forms of intelligence differ from machine (computer) intelligence?

Question #2: What are the cultural roots of privileging print over oral representations of experience? What are the limitations of print? Of orally based cultural narratives and thinking?

Question #3: In what ways does digitally based communication misrepresent the cultural and natural ecologies? What are its advantages in learning about the behavior of natural ecologies? What aspects of culture cannot be digitized? What are the advantages and disadvantages in digital representations of the behavior of cultural ecologies? Do digital representations also encode the cultural assumptions and silences of the computer scientist and programmers?

Question #4: What are examples of efforts within the culture of computer scientists to avoid being entrapped by the cultural assumptions that underlie the paradigm that the industrial/individualistic/consumer-dependent lifestyle continues to be based upon?

Question #5: What is lost when the intergenerational knowledge, skills, and mentoring relationships that underlie the culturally diverse cultural commons are no longer passed on through face-to-face relationships, but instead as part of the digitally stored data and information that can be accessed on the Internet? Can mentoring relationships be digitized and represented in print or in a video without a loss of what is unique about mentoring?

Question #6: What will be the impact on different cultures when their moral and identity-shaping narratives are no longer passed on through face-to-face relationships and ceremonies, but instead are available on the Internet and through video games and other entertainment venues?

Question #7: As computer scientists and futurist proponents of a digitally connected world monoculture, assume that Darwin's theory of evolution provides the best explanation of the role of digital technologies in this cultural transforming process, does the key Darwinian insight about the role of natural selection have implications for determining the moral values that should guide behaviors in the world's diversity of cultures? Does the Social Darwinian phrase "survival of the fittest" become the basis for determining which moral values should be carried forward? Does the phrase "better adapted" mean the same thing as "survival of the fittest"?

Question #8: What are the deep cultural assumptions that are reinforced by the growing reliance upon digital technologies, and do these cultural assumptions—including the reliance upon the theory of evolution—support the political ideologies of market liberalism and libertarianism?

Question #9: The rate of technological and environmental change indicates that the world's diverse cultures are facing a critical tipping point: namely, whether the dominant forces that promote change will also be able to contribute to conserving the different traditions of social justice, the gift economy of the cultural commons that enables people to live more community-centered and less money-dependent lives, and the natural systems that are under increasing stress from industrial chemicals and the loss of habitats. Do digital technologies provide for conserving the life-sustaining traditions of different cultural and natural ecologies, or do they leave the process of conserving the biological and cultural heritage up to the judgment of individuals? Do these technologies enable educating current and future generations in ways that are different from oral and written narratives about the traditions of injustice and exploitation that should be changed—or is this to be left to the self-discovery of what is on the Internet?

Question #10: Do computer scientists and the corporations have the right to transform the knowledge and moral systems of different cultures without engaging these diverse populations in a discussion of the changes that are about to eliminate many of their cultural traditions? Does face-to-face democracy have a place in the emerging world order envisioned by computer scientists? Does face-to-face democracy involve a different form of accountability than what is experienced in computer-mediated voting?

Question #11: Moral values and spirituality are dimensions of human experience that have developed over the centuries. Are there examples of how these aspects of human experience, which vary widely in different cultures, have contributed to addressing social justice issues and to living less environmentally destructive lives? Most proponents of the digital revolution either assume that humans will be replaced by super-intelligent machines that will make moral decisions obsolete or that all moral decisions relating to human and environmental relationships will be settled by access to massive amounts of data. Will these technological developments diminish human experience or are moral values and spiritual awareness of being connected with the land examples of an earlier stage in the evolutionary process that can now be left behind, as E. O. Wilson suggests in *Consilience: The Unity of Knowledge* (1998)?

Question #12: Given that most people in the West continue to think that technologies are culturally neutral, and that the value system of the person using the technology determines whether it is a positive or destructive force, how would you introduce a conceptual understanding that technologies are not culturally neutral? What examples would you give that would be difficult to ignore?

Question # 13: After viewing the talk by Michael Huesemann (see video at www.youtube.com/watch?v=GQ_kci8n09M), ask what strategies would be most effective in persuading computer scientists and their corporate supporters that the introduction of new digital technologies should be responsive to the democratic process?

The Double Binds We Face When Introducing Educational Reforms

Basically, the double bind we face is that what is being understood as the tipping point in the development of new technologies, one that will enable billions of people to experience the progress inherent in the emerging digital culture, also involves the loss of those aspects of the world's cultures that cannot be reduced to data and stored in the new cloud technologies and in massive servers. Part of the problem is that western universities, given the smorgasbord of degrees and courses they offer, do not require all students to learn about their own taken-for-granted culture or that of other cultures. The printed accounts of what too often turns out to be a fragmented account of the student's culture is filled with biases and questionable interpretations that seldom leave the realm of abstractions. The student's education is further jeopardized by how computer scientists and other faculty must be relied upon to promote a more complex and balanced understanding of the cultural transforming nature of different technologies. In most instances, they both lack the most rudimentary understanding of the cultural traditions displaced by different technologies. This has been one of the trade-offs resulting from the gains achieved by highly specialized approaches to inquiry, research, and teaching.

So the question becomes: how can faculty (including classroom teachers) introduce students to the cultural transforming nature of technologies, to how different technologies diminish human experience and potential, as well as its constructive and destructive role in addressing the cultural roots of the ecological crisis? That is, how can they overcome the limitations of their own education and the pressures within their discipline to not stray across intellectual boundaries? Added to the problem is that few faculty will have read any of the major books that have focused on the cultural transforming, and thus political, nature of technologies.

As the rate of environmental changes accelerates, and as the digital revolution leads religious fundamentalists (whom we call terrorists) to recognize that, in addition to providing new weapons systems, the digital culture is also undermining their way of life and belief systems, the world is going to become more dangerous for everyone. Faculty may then be more likely to recognize that the cultural, political, and ecological implications of technologies, especially the new digital technologies, can no longer be ignored, and that students should be engaged in discussing the most pressing issues. The question then becomes: how do they start and what should they read in order to avoid falling into the trap of what is now considered a politically correct pedagogical practice of asking students what they want to learn about technologies. I saw this approach being taken in a graduate class. The books suggested by the students required a depth of background understanding that all of the students lacked. And the books selected by individual students were unrelated to any organizing conceptual frameworks that could be applied to a broader set of issues. It was a total waste of time for everyone!

The ideal situation would be for faculty to read the major works of Eric Havelock, Jack Goody, and Walter Ong on the differences between orality and the technology of print. It would also be important to read Michael Reddy's piece on how the conduit view of language hides awareness of the metaphorical nature and thus the history of words, as well as Bateson's insights about the metaphorical nature of the language/thought connections, and the recursive epistemologies of the West. Bateson is also important for under-standing the nature of ecological intelligence. My book, *The Way Forward: Educational Reforms that Focus on the Cultural Commons and the Linguistic Roots of the Ecological/Cultural Crises* (2012), expands on Bateson's understanding of ecological intelligence by examining how print undermines the exercise of ecological intelligence. This book also makes the argument that revitalizing the cultural commons represents a life and community-sustaining alternative to one of the primary goals of the digital revolution, which is to reduce the need for workers and thus to force people further into poverty.

Given the reality of faculty continuing to work within the conceptual boundaries of their discipline, and only having a limited amount of time and energy to learn about the basic technology issues that should be intro-duced if students are to become minimally competent in challenging the inappropriate uses of digital technologies, I would like to suggest that they read the following: Jacques Ellul, *The Technological Society* (1964); Don Ihde, *Technics and Praxis* (1979); Langdon Winner, *Autonomous Technology* (1977) and *The Whale and the Reactor: A Search for Limits in an Age of High Technol-ogy* (1986); and Lewis Mumford, *The Pentagon of Power* (1970)—volume 11 of *The Myth of the Machine*, with special attention to Chapter 12. Reading any one of these books will disclose just how uninformed the computer scientists are about the world they and their capitalist supporters are work-ing to dominate.

There is another approach to consider that does not rely upon the faculty member taking on the sole responsibility for developing the curriculum for a course. Assuming that a group of faculty become concerned about the cultural and ecological implications of the digital revolution envisaged by computer scientists, an alternative would be to take an interdisciplinary approach. Just as with an interdisciplinary approach to introducing students to the technological, economic, and social forces that led over time to the enclosure (e.g. monetizing, privatizing, and commodifying) of the cultural commons—that is, intergenerational knowledge and skills in the areas of music, healing practices, craft knowledge, food production and preparation, and narratives—an interdisciplinary approach to developing a curriculum that introduces students to the cultural transforming nature of technology could also have the same breadth of historical and cross-cultural perspectives. Equally important, this approach would provide an in-depth study of specific examples of the unanticipated consequences of adopting different technolo-gies, of those who gained and those who became further impoverished, and of what changed the direction of cultural development. The interdisciplinary approach would also have another advantage, as it would provide a deeper historical perspective on specific cultural examples.

Ideally, a combination of the two approaches would help to overcome the current silences in most university curricula about one of the dominant influ-ences on everyday life—and now on whether digitally based information and thinking will enable students to understand the cultural/linguistic roots of the ecological crisis. The long-held idea that technology becomes a constructive or destructive force, depending upon the ideas and values of the people who promote its use, is partly correct. But only partly. Its cultural amplification and reduction characteristics, which need to be understood within the contexts of different cultural knowledge and value systems and their environmental contexts, also need to be understood—and not just assumed to be a progressive force driven by Nature's process of natural selection.

References

Bateson, G. 1972. *Steps to an Ecology of Mind.* New York: Ballantine Press.

Bowers, C. 2000. *Let Them Eat Data: How Computers Affect Cultural Diversity, and the Prospects of Ecological Sustainability.* Athens, GA: University of Georgia Press.

_____. 2011. *University Reform in an Era of Global Warming.* Eugene, OR: Eco-Justice Press.

_____. 2012. *The Way Forward: Educational Reforms that Focus on the Cultural Commons, and the Linguistic Roots of the Ecological/Cultural Crises.* Eugene, OR: Eco-Justice Press.

Ellul, J. 1964. *The Technological Society.* New York: Vintage.

Goodman, P. 1961. *Growing Up Absurd: Problems of Youth in Organized Systems.* New York: Random House.

Goody, J. 1977. *The Domestication of the Savage Mind.* Cambridge: University of Cambridge Press.

Hardin, G. 1968. The tragedy of the commons. *Science,* 162(3859), 1243–1248.

Havelock, E. 1986. *The Muse Learns to Write: Reflections on Orality and Literacy from Antiquity to the Present.* New Haven: Yale University Press.

Ihde, D. 1979. *Technics and Praxis.* Dordrecht: D. Reidel Publishing.

Illich, I. 1971. *Deschooling Society.* New York: Harper and Row.

Illich, I. 1973. *Tools of Conviviality.* New York: Harper and Row.

Illich, I. and Sanders, B. 1988. *The Alphabetization of the Popular Mind.* Berkeley, CA: North Point Press.

Lakoff, G. and Johnson, M. 1980. *Metaphors We Live By.* Chicago, IL: University of Chicago Press.

_____. 1999. *Philosophy in the Flesh: The Embodied Mind and Its Challenge to Western Thought.* New York: Basic Books.

Mander, J. 2012. *The Capitalism Papers: Fatal Flaws in an Obsolete System.* Berkeley, CA: Counterpoint.

Mumford, L. 1970. *The Pentagon of Power: The Myth of the Machine.* New York: Harcourt, Brace Javanovich.

Ong, W. 1982. *Orality and Literacy: The Technologizing of the Word.* New York: Routledge.

Reddy, M. 1979. "Conduit Metaphor: A Case of Frame Conflict in Our Language about Language." In *Metaphor and Thought,* A. Ortony ed. Cambridge: Cambridge University Press.

Wilson, E. O. 1989. *Consilience: The Unity of Knowledge.* New York: Alfred A. Knopf.

Winner, L. 1977. *Autonomous Technology Technics-out-of-Control as a Theme in Political Thought.* Boston, MA: M.I.T. Press.

Winner, L. 1986. *The Whale and the Reactor: A Search for Limits in an Age of High Technology.* Chicago, IL: University of Chicago Press.

14 The Digital Revolution and the Unrecognized Problem of Linguistic Colonization

A combination of forces are preventing a wider awareness of the ways in which the digital revolution represents a colonizing force in the world today. While its promoters claim it to be a progressive and modernizing force, an examination of just one of its colonizing characteristics reveals that it is undermining the world's diversity of intergenerational knowledge of how to live less consumer-dependent and thus less environmentally destructive lives. The failure of universities to promote an understanding of a number of language issues that have a direct impact on exercising ecological intelligence also contributes to the widespread failure of both computer scientists and their supporters, as well as the general public, to understand how the digital revolution is changing the world's cultures in ecologically unsustainable ways. The silences on the part of universities make them complicit in another feature of the digital revolution: namely, that computer scientists, programmers, and the growing army of technological entrepreneurs do not understand the cultures into which their technologies are being introduced and thus are unaware of their cultural and environmentally destructive nature. What goes unrecognized is that the same progress-at-all-cost, individually centered ideology that provided conceptual guidance and moral legitimacy to the first industrial revolution now guides the digital revolution.

The complex network of cultural and natural ecologies that support everyday life, which are characterized as emergent, relational and interdependent, are both sustained and undermined by the multiple forms of communication integral to all ecological systems. In terms of the world's diversity of cultural ecologies, the communication takes many forms—ranging from the spoken and printed word to learning from what is being communicated by changes in the natural environment. In the West, the primary forms of communication that pass forward the traditions of a culture and sustain what people take to be "reality" are the printed and spoken word. While there are other patterns of communication experienced as taken-for-granted sources of information and meaning, and which are marginalized by what cannot be digitized, the focus here will be on how the digital revolution undermines the importance of oral communication while reinforcing the more problematic characteristics of print.

What is Problematic about Print and Data-Based Cultural Storage, Thinking, and Communication

Print has many important—indeed, essential—uses, but it is also limited in representing the primary characteristic of all living cultural and natural ecologies. These emergent and relational life-forming and sustaining processes have been misunderstood by the West's philosophers and social theorists who represented the world as made up of autonomous entities such as things, abstract ideas, individuals, events, plants, animals, and so forth. That is, because the philosophers privileged abstract thinking, they left a legacy of ignoring the fact that nothing exists free of relationships within the larger cultural and natural ecologies. As the digital revolution relies upon the printed word and other abstract systems of representation, it reproduces key features of the printed word. Print, even when used by the most gifted writer, can never fully represent the emergent, relational, and interdependent nature of an experience. For example, print can never fully represent the experience of watching a wave crashing against the rocks or trying to engage a libertarian/market liberal in a discussion about the connections between putting billions of tons of carbon dioxide into the atmosphere and the growing rate of acidification of the world's oceans.

Print is useful for storing and communicating information and data, but it also contributes to the tradition of abstract thinking that has been such a powerful reality-shaping force in the West. That is, what is encoded in print immediately becomes dated (given the emergent nature of reality), cannot provide a full account of contexts, reinforces the misconception that there are ideas, things, individuals, and so forth that are autonomous (print, like English nouns, is inadequate in communicating ongoing relationships), and reduces the importance of learning from all the senses and giving special attention to local contexts. In addition, print fosters a taken-for-granted acceptance of the surface knowledge that print represents. This surface knowledge, given the dynamic contexts that print cannot represent except as data, information, and other abstractions, leads to a culture of abstract and surface thinkers. Evidence of this can be seen in the print-based rational process of most Western philosophers who were ethnocentric thinkers, and whose theories seldom addressed cultural issues except when providing a culturally uninformed explanation of the nature of private property and free markets, and why rational thought is superior to face-to-face experience and narratives. This legacy can be seen in how much of daily (especially political) discourse relies upon words such as "freedom," "individualism," "technology," "data," "intelligence", "progress," "competition," "growth," "conservatism," and so forth. Their abstract use can be seen in how actual cultural and natural contexts are ignored. Thus, so-called "conservatives" are not held accountable for the traditions of community self-reliance they undermine in order to expand markets and profits. And abstractions such as "individualism" and "tradition" do not take account of the different linguistic/cultural ecologies that influence taken-for-granted patterns of thinking and values within different cultural contexts.

As data is acquired through various approaches to observing and measuring behaviors of natural and cultural processes, it shares the same abstracting limitations as print. Instead of considering the deeper implications of the surface nature of what data actually represents, the immediate concern of the technocratic/market-oriented mindset is to interpret its importance in terms of how it can be used to achieve greater efficiencies or to solve a problem that reflects the interests of the person or organization that collects the data. In short, data is unable to fully represent the emergent, relational, and interdependent nature of the cultural and natural ecologies, such as the ecology of workers and the ecology of those living below the poverty line. Data is unable to account for the worker's inner experience of being replaced by a robot and the emerging network of relationships that must be negotiated if food and shelter are to be available. Unfortunately, data which is inherently an abstraction has become high-status knowledge, with the more complex and context-based knowledge that comes from lived experience becoming represented as inferior to data as it is not "objective." Yet the word "objective" is another abstract metaphor that precludes considering the culturally influenced ways of knowing and values that determine what is to be observed and measured—and how the supposed "objective data" is to be interpreted by people who are seldom aware of the deep cultural assumptions they take for granted.

This critique of how the digital revolution promotes abstract thinking should not be interpreted as suggesting it has not led to many benefits. The printed word appearing on the computer screen and the data-driven models and decisions have led to important gains in the quality of life, and in learning about the changes occurring in natural systems. The problem is the lack of a balanced understanding of the beneficial and destructive uses of digital technologies. In addition to how the computer scientists and cowboy capitalists think primarily in terms of progress, profits, and of bringing cultural and natural processes under the control of the Internet of Everything (which is a code phrase for bringing all aspect of daily life under total surveillance), it is important to identify other ways in which the digital revolution is undermining the prospects of an ecologically sustainable future.

The Conduit View of Language and the Loss of Awareness of the Metaphorical Nature of Language

In 1979, Michael Reddy published a paper critiquing what he referred to as the conduit view of language. This view of language, or more accurately, how we use language in what we assume to be a sender/receiver process of communicating data, information, and rational ideas, has been central to a number of myths perpetuated at all levels of education—and now by the digital revolution. The conduit view of language is essential to maintaining the myth of objective knowledge and that the rational process is free of cultural influences—two criteria that have importance in colonizing other cultures. What the conduit view of language marginalizes is one of the most important characteristics of written and spoken language that have especially important implications for addressing the cultural roots of the ecological crisis. That is, it undermines awareness that words have a cultural history, and that most words are metaphors whose meanings were framed by analogs settled upon by earlier generations of Western thinkers who were unaware of environmental limits—including the silences and prejudices of their era. We now recognize how the meaning of the word "woman" was framed by the prejudices and other misconceptions of earlier eras, and how nature was viewed as dangerous and in need of being brought under human control. Most of our vocabulary are metaphors, including words such as "property," "traditions," "free markets" and so forth, that reproduce the earlier constituted analogs that become the basis of thinking of succeeding generations—which leads to the problem Einstein identified when he warned against relying upon the same mindset to fix the problems that created it. The biographical variations in people's lifestyles, including awareness of the discrepancies between how the inherited metaphorical vocabulary fails to take account of the emergent realities of everyday life, may lead to old metaphors being challenged and reframed in terms of ecologically and culturally informed analogs. "Wilderness" and "woman" now have different meanings than in earlier eras. Given the ecological crisis, we now need to identify ecologically and culturally informed analogs for the meaning of such metaphors as "intelligence," "tradition," and "progress" (Bowers 2011, pp. 69–92).

What appears on the computer screen, whether as a YouTube presentation, information and data on a website, a computer-mediated curriculum unit, or an email, will not include the warning that the words appearing on the screen or heard on a cell phone have a history, and that they too often carry forward the meanings framed by the analogs settled upon in the distant past. That is, the digital revolution reinforces the conduit view of language and thus carries forward the taken-for-granted assumption that words refer to real events, knowledge of objects, behaviors, data, information—with no references to their linguistic histories that encode different cultural ways of knowing. Printed words, because they share the limitations mentioned above, further reinforce abstract thinking. It is the taken-for-granted acceptance of abstract thinking that leads computer scientists, programmers, the growing army of technological entrepreneurs, as well as the general public mesmerized by digital technologies, to overlook the cultural traditions that are being undermined.

Other Aspects of Culture Not Understood by Computer Scientists and Their Supporters

The scope of the deepening ecological crisis—which includes the growing acidification of the world's oceans, extreme climate changes, droughts and wildfires, loss of species and habitats, and the poisoning of natural systems with the millions of tons of toxic chemicals—needs to be taken into account in terms of how progress is understood. That the digital revolution is the driving force in expanding markets and thus consumerism that will deepen the ecological crisis, and in introducing other life-altering changes such as replacing workers with computer-driven systems, it is necessary to consider other digitally driven changes that further undermine the prospects of an ecologically sustainable future. For example, what is not recognized by the proponents of the digital revolution is that the world's diversity of cultural commons represent alternatives to a consumer-dependent existence that is environmentally destructive. People are beginning to turn to these largely non-monetized community-centered cultural commons as they recognize how the industrial/consumer-dependent lifestyle cannot be sustained by natural systems now in rapid decline.

The cultural commons vary from culture to culture, but share common features. The main one is that the intergenerational knowledge and skills that carry forward traditions of mutual sharing in growing and sharing food, healing practices, ceremonies and narratives that carry forward the moral templates that guide human/nature relationships, creative arts and craft skills, games, and knowledge of local ecosystems, are intergenerationally renewed through face-to-face communication. That is, the oral traditions are essential to the processes of mentoring and to the formation of personal identities and values. It is the oral traditions, rather than print, that connect the current generations to the knowledge and skills that have been refined over generations of how to live in mutually supportive and non-commoditized relationships (Bowers, 2012).

The digital revolution undermines the oral traditions essential to the intergenerational renewal of the cultural commons by reinforcing the West's long history of privileging print and other abstract systems of representation as having higher status. The long-standing bias against oral traditions can be seen in how the word "illiterate" carries the connotation of backwardness and ignorance. The more immediate impact of the digital revolution now being experienced in cultures that are still predominately based on orally shared intergenerational knowledge is that their youth are being indoctrinated into thinking that the Internet provides access to the excitement and information necessary for a modern existence. This is leading to the alienation between generations, and thus to the digital generation of youth failing to learn the knowledge and skills that enabled the older generations to live in mutually supportive relationships within the limits and possibilities of their bioregions. Contrary to current misconceptions, the revitalization of the cultural commons does not involve returning to the lifestyle of earlier centuries, but rather learning the current largely non-monetized traditions being carried forward in every community. The current cultural commons practices, which differ significantly between ethnic groups, provide alternatives to a future where employment will become even more limited as the vision of progress that drives the digital revolution promotes replacing workers with robots and computer systems. A further way in which the digital revolution undermines the cultural commons existing in every community across America can also be seen in how much time is spent playing video games, texting, surfing the Internet, and communicating on social networks. These activities lead to a further disregard for learning about the differences between ecologically sustainable and unsustainable traditions—including how the political economy of the local cultural commons provides for the discovery of personal talents and skills that are denied in consumer-dependent relationships.

It is important to recognize that digital technologies are used within different cultural commons activities that range from the local farmers' markets (now having doubled in number in the last few years), to communicating with mentors and scheduling events where intergenerational knowledge and skills are shared, to learning about the changes in natural systems that need to be kept in focus. Again, it's a question of balance and of knowing the appropriate and inappropriate uses of digital technologies—just as it is a matter of recognizing when to rely upon the printed word and when to recognize when it reproduces the misconceptions and silences of earlier generations.

A point made at the outset needs to be emphasized in judging whether the digital revolution is the progressive force that many now claim it to be. If we consider the specialized education of the computer scientists, as well as the print-based and thus abstract education received in most areas of higher education—including the current idea that students should decide what they want to learn—we find that increasing numbers of graduates encounter the same surface knowledge of their own culture as encountered by people whose education is limited to public schools. That is, they do not have an in-depth understanding of the cultural assumptions underlying their own culture, and how the metaphorical language they take for granted reproduces these assumptions.

The higher students go in the formal educational process, the more they are indoctrinated to accept the misconceptions of the Enlightenment thinkers of the 17th century. That is, the emphasis in higher education on progress, innovations, new ideas and values, and abstract thinking, leads to viewing traditions, including those that sustain the cultural commons, as impediments to progress and innovation. But this Enlightenment way of understanding traditions, which the digital revolution reinforces, represents yet another set of traditions that have deeply problematic implications. As Edward Shils (1981) noted, the anti-tradition traditions of scientists, technologists, capitalists, and proponents of critical thinking view the emancipation from all traditions as ensuring a prosperous and progressive future. Unfortunately, they have not learned to consider which traditions contribute to an ecologically sustainable future and carry forward important achievements from the past—such as civil liberties, gains in social justice, and in other areas of the cultural commons.

This lack of awareness on the part of the promoters of the digital revolution of the traditions that need be intergenerationally renewed should be a major concern. Yet there are few people who are protesting the loss of a number of important traditions due to the technologies created by computer scientists and their promoters. These traditions include the loss of privacy, personal security now so widely compromised by hackers, safeguards from foreign cyber-attacks, expectations that employment will survive automation, non-militarized police forces still under civil control, and the knowledge that one's behaviors are not being monetized by corporations selling the data to governments and businesses—with the latter now adjusting their online prices in ways that take account of one's economic circumstances. People with a strong sense of social and eco-justice also valued the tradition of resisting the colonization of other cultures, but this tradition has now yielded to the idea that progress dictates the global spread of the digital revolution.

Given the rate of environmental degradation and the loss of the intergenerational knowledge and skills essential to revitalizing the cultural commons, it would seem that conserving species, habitats, and the diversity of cultural traditions that have a smaller adverse ecological impact would become a primary focus of our educated elites. However, in reading the computer futurist writers such as Eric Schmidt, Ray Kurzweil, Peter Diamandis, among others, there is no mention of the ecological/cultural crises—only the need for experts to replace human capabilities with computer systems.

References

Bowers, C. 2011. *University Reform in an Era of Global Warming*. Eugene, OR.: The Eco-Justice Press.

_____. 2012. *The Way Forward: Educational Reforms that Focus on the Cultural Commons and the Linguistic Roots of the Ecological Crisis*. Eugene, OR.: The Eco-Justice Press.

Reddy, M. 1979. "The Conduit Metaphor—A Case of Frame Conflict in Our Language About Language" (pp. 285–323). In *Metaphor and Thought*, edited by Andrew Ortony. Cambridge, UK: University of Cambridge Press.

Shils, E. 1981. *Tradition*. Chicago: University of Chicago Press.

15 Has the Authority of Data Consigned Wisdom to the Junk-Heap of History?

The digital revolution is producing profound changes in the world's cultures. What is unique about these many changes is that they are embraced by many people ranging from scientists, business women and men, educators, average citizens, and just about everybody else who values convenience, instantaneousness, multiple forms of empowerment, and the ability to escape from face-to-face relationships into the seemingly boundless world of data. The combination of surveillance technologies—connectivity, multiple monitoring systems, and storage—bring all aspects of the natural world as well as cultural life under the new God of capitalism and rational data-based decision making. Quickly disappearing from human memory are the various mythologically centered Gods that provided an integrated and morally coherent world view, and were renewed through rituals and narratives—and in many instances prescribed the punishment fitted to different moral transgressions. There are extreme reactionary movements now resisting how these ancient belief systems are being replaced by the new God of data and rational decision making. This only strengthens the efforts of the elites guiding this new religion to marginalize the voices of those who resist being converted to the modern, secular, and data-based understanding of the emerging world.

While this new God, and its new priesthood, have not totally displaced the God of the Old and New Testaments, its emphasis on the authority of data is bringing about fundamental changes in the vocabulary of various cultures used in the past to carry forward their wisdom traditions. For those closest to the center of this digital revolution, the word wisdom is seldom if ever used. When Bill Gates, an early prophet of this new religion, is purported to have claimed that we need to recover wisdom, few people would have understood what the word previously referred to, and what it might mean in the modern world where data is understood as eliminating subjective judgments and interpretations based on archaic moral narratives. The vision of 17th and 18th-century Enlightenment philosophers is at last being realized by how computer scientists are now putting decision making on an objective basis that relies upon data. The authority of objective knowledge, information, and now data even transcends the murky realm of politics that is too often influenced by memory and values derived from the pre-scientific world of ancient religious narratives.

Those who refuse to recognize the authority of data are still looking through a glass darkly. This archaic mindset leads to raising questions that cannot objectively answered—such as the differences between wisdom and data. Taking seriously the differences would require entering a realm already colonized by the followers of the scientific method who have demonstrated the power to predict the behavior of particles moving through space. The astonishing achievements of scientists suggest that we do not need to understand wisdom. What can wisdom help us understand if science has given us the ability to land men on the moon, and to genetically alter the basis of different forms of life? Besides, understanding wisdom first requires understanding the diversity of how humans have understood the nature and sources of wisdom. What citizen of the digital age can take time away from keeping up with the Tweets, cell phone and e-mail messages from friends and employers who expect their employees to be continually connected? And who is interested in entering the rabbit hole of human history chronicled by the winners, and who is genuinely concerned that the abstract world of data misrepresents the emergent, relational, and co-dependent life-sustaining processes in the natural and cultural ecologies within which we live? Isn't it enough that data can be used to reveal trend lines in profits, the expansion or reduction in crime rates, and the rate of acidification of the world's oceans?

There is no question that the abstract world of data is genuinely useful even when it represents a formulaic response that hides the many limitations both in what data is supposed to represent and in the moral issues seldom recognized in the political decisions surrounding their use. As the above sentences suggest, moving outside the certainties of an objective and measurable world also requires understanding that one's thinking is based on a culturally specific taken-for-granted ontology. This culturally constructed world is assumed to be composed of fixed entities such as autonomous individuals, abstract ideas and values that are both objective and have universal status, and the life force called progress is assumed to be like a road sign pointing the direction the rest of the world is to follow.

Recognizing the conceptual foundations of one's own taken-for-granted ontology seems like an unnecessary detour when data is so easy to understand—understanding an ontology that avoids colonization because it recognizes that all life-forming and sustaining processes, across the entire range of natural and cultural systems, are emergent, relational, co-dependent, and sustained by different ecologies of communication. This would require more than historical knowledge. That is, it would require a knowledge of other cultures—especially those that recognized that the emergent, relational, and co-dependent world within which they live are the basis of sustainable forms of ecological intelligence. For the typical citizen of the emerging digital culture, this effort would seem to be a waste of time as a new class of experts, the data scientists, as well as algorithms (and the computer scientists, programmers, and engineers working behind the scenes to create autonomous algorithms) possess the form of intelligence that easily turns data into decisions.

The Taken-for-Granted Ontology of the World of Data

Before discussing why wisdom is needed in a world that increasingly relies on data-based decision making, as well as how data misrepresents the world we live in, it needs to be acknowledged that for all its limitations data is useful in providing a better understanding of patterns, trend lines, causal relationships, rates of change, and changes in effectiveness and efficiency. It provides, in many instances, a more accurate account of the behavior of social and natural systems that might otherwise be misrepresented by a lack of close attention, and by efforts to hide the shortcomings in human behavior. For example, without data we have to rely upon conjectures and traditional misconceptions about the behavior of marine ecosystems. Data provides a more accurate understanding of how many sharks are being killed each year in order to satisfy a traditional cultural preference for shark fin soup. Similarly, data provides a more accurate understanding of how fraudulent Medicare and Social Security claims are distorting the national budget. Data is also useful in providing an expanded understanding of other cultural patterns of behavior relating to gender and racial discrimination, and so forth.

In spite of its many uses, data, like the scientific method, tells us "what is" within a limited context. It does not tell us how we "ought" to respond to the issues and problems revealed by the "what is" information. In order to understand the limitations of data and the role of cultural influences that are largely ignored due to an over-estimation of the authority that has been conferred on data, it is necessary to take account of the following:

1   Like print, taken-for-granted cultural assumptions influence both what is regarded as important to represent in the form of data, as well as the interpretation of how it is to be used. That is, while data is assumed to be objective, there is always a decision made by an individual or group about what is to be measured and represented as data. This decision is culturally influenced because the thinking and values of the decision makers are influenced by the languaging processes that tacitly reproduce earlier cultural ways of thinking and valuing. As data represents only a segment, like a snap-shot, of what is emergent, relational, and co-dependent within the larger ecological system, what it represents, like René Magritte's famous painting "Ceci n'est pas une pipe," is only a partial, abstract, and symbolic image—and not the pipe itself. In short, data is only a surface representation, and it encodes the assumptions that are at the front end, the initial decisions, of the data-gathering process.

2   How the data is interpreted is also a culturally influenced process. The mindset of the individual and group interpreting the data in terms of

what it means is always under the influence of the cultural assumptions that are taken for granted. For example, the environmental scientist brings a different set of assumptions and values to the process of interpreting the data than that of the data scientists working for a corporation or an office of education concerned with acquiring "objective" evidence of learning outcomes (to use the jargon). What the myth of objective data requires overlooking is the ecology of linguistic influences, the ecology of identities, and the ecology of interpretative and moral frameworks that are variously called an ideology, the scientific method, and the individual's critical rationality. The cultural/linguistic ecology that influences both ends of the data collecting and interpreting process are inescapable aspects of the interpreted world in which we live. That we can escape into a world of objective facts, data, and the printed world is a modern myth.

3   Because the surface and momentary measurement or observation of a phenomenon does not take into account its larger dynamic context, and because many westerners carry forward the Cartesian tradition of thinking of themselves as rational spectators of an inert, material, external world, data (as well as print) reinforces a basic ontological misconception about a world of permanent, fixed, and Platonic universal entities. Those who claim to have a rational and thus objective understanding of this abstract world too often possess power and authority over those who acknowledge they live in an impermanent and interpreted world.

4   The ideology that serves as an interpretive framework for determining the meaning and uses of data reinforces an instrumental moral framework that, with the exceptions of how environmentalists use data, serves the interests of market liberals who promote consumerism and the monetization of everyday life. This instrumental moral framework is supported by the cultural assumptions about the autonomy of the individual and the importance of progress in producing material wealth and in exploiting the environment.

It would not be too much of a stretch to claim that the dominant Anglo/European print-based culture, out of ignorance of its own modernizing assumptions, uses data as though it legitimates the decisions that lead to further economic progress. That is, data is being viewed as providing both an account of "what is" as well as what "ought to be." Actually, what is represented as data is too limited, and too much a reflection of the assumptions of the experts setting the data-gathering process in motion, to provide the moral guidelines for how it is to be used. The moral and instrumental guidelines are derived, instead, from the prevailing taken-for-granted ideology of the social groups seeking legitimation for their decisions. If this were recognized, the ideology of the group masking their policy decisions, including the justifications for replacing people with machines, might more often be challenged. But how many people have been educated to recognize how certain words in the vocabulary, such as "objective," "rational," "progress," "expert," "science," and now "data," are assumed to represent certainties that are beyond political debates? The irony is that when judged in terms of the past decisions of the ideologically driven groups who have relied on data to justify their economic and political agendas—in promoting technological innovations, in monetizing the cultural commons, in colonizing other cultures, and in educating the country's youth to equate success and happiness with climbing the pyramid of consumerism and wasteful living—data-based decision making has been both de-humanizing as well as ecologically destructive. Indeed, data has become the common currency shared within the interlocking surveillance technologies that are putting the country on the road to a techno-fascist future.

If current market and individually centered ideologies are accelerating environmental changes that are leading, as some scientists now claim, to the sixth extinction of life on this planet, then the question about the recovery of wisdom becomes not only more relevant, but more urgent. When we consider not only the wisdom traditions within different cultures, but also how these traditions were influenced by profoundly different cultural mythologies/epistemologies, centuries of learning how to encode their guiding moral frameworks in narratives, dance, and in every aspect of their cultural commons, as well as in their relationships with the natural world, the question arises as to whether a knowledge of the wisdom traditions of other cultures will lead to fundamental changes in the western mindset in time to avert the social chaos and ecological endgame that lies just decades ahead. In spite of my increasing doubts that the majority of academics and experts guiding a variety of innovative agendas will take seriously the challenge of basing decisions on wisdom rather than data, I will nevertheless identify a number of wisdom traditions that still guide human/nature relationships—and from which groups on the fringes of the mainstream individualistic, consumer-dependent, and profit-oriented culture are learning.

Two Ancient Relationally Oriented Wisdom Traditions: Buddhism and Confucianism

The fundamental differences between the cultural patterns reinforced by data-based storage, thinking, and communication can be seen by comparing the difference between what the Buddhists call the Path, and what they refer to as "wondering about"—which refers to the lifestyle that is not reflective and is continually influenced by outside forces and shifting subjective whims. The lifestyle of "wondering about" is exemplified in the West as a consumer-driven lifestyle and the many illnesses that accompany it. It is also reinforced by how the Internet reinforces change, short attention spans and memory, and an instrumental approach to information and data.

The Path, on the other hand, leads to a lifestyle of mindfulness and thus to a radical transformation in life's guiding principles. The names of the Path's eight steps include the following: 1) Right views; 2) Right Intent; 3) Right Speech; 4) Right Conduct; 5) Right Livelihood; 6) Right Effort; 7) Right Mindfulness; 8) Right Concentration (Smith, 1991, pp. 105–112). It is notable that possessing the right amount of data is not included as contributing to the path of mindfulness. As the behavioral and thought process associated with each of these steps is elaborated upon, it becomes clear that Buddhism is focused on the moral and spiritual dimensions of relationships as they are experienced in a constantly changing world. It is also clear that the Path requires a lifelong commitment, which differs radically from the short attention span and expectation of obtaining instantaneous results reinforced by cyberspace experiences. Perhaps more important in terms of the need to reduce the human impact on natural systems, the Path represents an alternative to the consumer-dependent lifestyle valued in the West. It is also important to note that different traditions of Buddhism are being taken seriously in the West, but not in sufficient numbers to have a real impact on the still growing influence of the digital revolution that supports the global expansion of the market system.

Confucianism, like Buddhism, is also a religion so deeply ingrained in daily cultural practices that it is understood more as the taken-for-granted reality of daily life. Its fivefold principles include the following: **Jen**, which "involves simultaneously a feeling of humanity toward others and respect for oneself, an indivisible sense of the dignity of life wherever it appears." **Chun tzu** highlights relationships that are the opposite of the competitive, petty, and ego-centered person. The person of Chun tzu puts others at ease and engages in what Martin Buber later referred to as I-Thou relationships and dialogue. **Li** is the quality that leads to doing things correctly—in the use of language, in avoiding extremes, in the correct ordering of relationships within the family and society. **Te** is the power of moral examples that attract the willing support of the people. **Wen** refers to the "arts of peace," specifically the power of the arts to transform human nature in ennobling ways (Smith, 1991, pp. 175–181). There is no mention of the importance of data in these life-guiding principles. But the digital revolution, which is central to economic growth in China and other cultures with a Confucian past, is having a transformative impact on the youth of these cultures.

A critical issue is whether the wisdom traditions of Buddhism and Confucianism will survive as the mindset of the youth of these cultures is being shaped by the westernizing mindset of the digital revolution. Both the relational wisdom of Buddhism and Confucianism were intergenerationally renewed though face-to-face communication, through mentoring, and through being aware that others took for granted these principles as moral imperatives. The spread of market forces, rising material standards of living, slick media images connecting consumerism with individual happiness, and the role of the digital revolution in expanding the economies of Asian countries and in westernizing their approaches to education all work against youth even being aware of these ancient wisdom traditions—except to view them as the old and pre-modern ways of their grandparents.

Ecologically Informed Wisdom Traditions that are Sources of Resistance to the Individually Centered, Consumer-Dependent, and Data-Based Culture

The ecologically informed wisdom traditions that stand in sharpest contrast to the western mindset that is making a cult of data-based decisions, and that have the most relevance for learning how to live less environmentally destructive lives, represent the achievements of many of the world's indigenous cultures. Their wisdom was not acquired from abstract thinkers such as western philosophers, nor was it acquired from books and from data. Rather, it was acquired from living in one place over hundreds or even thousands of years, giving close attention to the cycles and patterns of interdependencies of life in the natural world, using myths as repositories of practical ecologically informed knowledge, narratives, and ceremonies that connected the generations in webs of meaning, rituals around food and healing practices, and renewing the knowledge and moral insights learned by previous generations by taking account of the ongoing changes in the local bioregion. What seems common to these wisdom traditions is that unlike the mythic account of "man's" fall in the *Book of Genesis*, and the injunction to name and subordinate the plants and animals to human will, they learned from the nature itself; that is, the "Garden of Eden," to stay with that metaphor. That is, rather than escaping from the Garden by creating a human-centered world of moral and conceptual dichotomies and categories, the indigenous cultures engaged in what is today known as biomimicry, which shows up in their metaphorical language and knowledge of local ecosystems.

Giving close attention to the information flowing within and between the natural systems, such as how the behavior of animals, even that of the tree, anticipates the severity of the coming winter, fosters reliance upon the exercise of ecological intelligence. Awareness of the interconnected patterns in a world of impermanence, as well as awareness that adapting how to meet human needs in ways that take account of these changing patterns is essential to sustaining life within the biotic/human community, are profoundly different from surface, abstract, snap-shot images we call data. There is no sense of the sacred in the world reproduced as data, and there is no awareness of an inclusive spirituality. Without a sense of the sacred and an inclusive spirituality, everything becomes possible to the mindset that reduces the cultural and natural ecologies to the data, including destroying forests, mountains, streams and rivers, and their multiple animal inhabitants if it leads to more profits and human conveniences.

The mythic thinking of the peoples who have inhabited the Andes for centuries, and whose understanding of Pachamama as the force that nurtures humans as humans nurture nature, has led to one of the world's mega-diversities of edible plants. It also represents many of the elements of wisdom shared by other indigenous cultures. As explained by Grimaldo Rengifo Vasquez (1998, p. 97):

> In the Andean world everything is alive and important; nothing is inert and nothing is superfluous. The very stone is alive, it speaks and the peasant converses with it as person to person. It is not that the peasant extends the notion of a person to the stone (which is generally understood as 'personification') but rather that, for the peasant, the stone is alive—possessing the attributes of the *runa* and vice versa.
>
> In the Andean context we cannot speak either of the inanimate as opposed to the animate, or of the essential as opposed to the contingent. The whole *Pacha* is a community of interconnected living beings, in which man and water are as important and alive as are the *huacas* (deities) and the wind in terms of the regeneration of life.

During my visit to Cajamarca, the site of Pizarro's capture and execution of Atahualpa, the sovereign emperor of the Inca Empire, my western consciousness was opened up to how the stone could be understood as being alive, and an active participant in the information networks that connected all forms of life in the bioregion to the cosmos. My western consciousness, oriented toward actions that increased efficiency and a humanly controlled world, led to wondering why the stones were not used as boundary markers as in England, France, and other western countries. Instead they lay scattered across the field. Following the advice to pick up a stone, I found how its surface appearance indicates the level of moisture in the soil—which is vital information for the farmer to understand. The number of eggs a bird lays, the number of animals in a herd—even the condition of their fur, and so forth, are signs of the current and forthcoming patterns operating in the regeneration of life. In effect, the wisdom carried forward from earlier centuries among the Andean peoples is that everything communicates, everything is part of the same spiritual and moral universe, and that these cycles of interdependence should not be broken. But they now are being broken as western extraction industries are tearing up the earth for oil, gold, and other resources needed to produce the throw-away, data-driven culture of the West.

The wisdom of the Aboriginal peoples that mapped and storied what we now call Australia for 40 to 50 thousand years also avoids the anthropocentrism of the tribal cultures that eventually put their narratives in print that we now read as the *Book of Genesis*. As recounted by Robert Lawlor in the *Voices of the First Day: Awakening in the Aboriginal Dreamtime* (1991), their cosmology was also the basis of their moral order, provided the wisdom that guided their uses of technologies, and resulted in a level of ecological intelligence that far surpassed the Anglo culture that invaded the land and set out to westernize them. Lawlor summarizes the wisdom that was integral to their cosmology in the following way:

> All creatures—from stars to humans to insects—share in the conscious-ness of the primary creative force, and each, in its own ways, mirrors a form of that consciousness. In this sense the Dreamtime stories perpetu-ate a unified worldview. This unity compelled the Aborigines to respect and adore the earth as if it were a book imprinted with the mystery of the original creation. The goal of life was to preserve the earth as much as possible, in its initial purity. The subjugation and domestication of plants and animals and all the other manipulation and exploitation of the natural world—the basis of Western civilization and 'progress'—were antithetical to the sense of a common consciousness and origin shared by every creature and equally with the creators. To exploit this integrated world is to do the same to oneself.
>
> (Lawlor, 1991, p. 17)

The cosmologies of the Quechua, the Australian Aborigine, as well as many other indigenous cultures, recognized a sacred and thus moral order that was (and is) profoundly different from the instructions in the *Book of Genesis* for man to name the creatures of God's creation and to take control of them. What is often not recognized is that the *Bible* was written by a tribal culture dedicated to a cosmology and moral order centered on a monotheistic God. The surrounding cultures that understood all forms of life as sacred and ani-mated by different spirits, and thus as participants in the same spiritual uni-verse, were regarded as challenging the one true God. The irony is that these first indigenous cultures were initially pursuing the path leading to ecological wisdom, while the author of the *Book of Genesis* (believed to be Moses) was laying the conceptual and moral foundations for the anthropocentic culture of the West, which would later become the foundation for the industrial and capitalist exploitation of nature. This anthropocentric cosmology, as well as early Biblical injunctions to take control of the earth and to multiply (both now contributing to the ecological catastrophe we are now entering), con-tinues as the basis of today's emphasis on progress that now relies so heavily on data-based decisions.

The youth in these indigenous cultures, from the Haida, Dene, Inuit, to thousands of other indigenous cultures spread around the world, are now caught between their ancient ecologically informed sources of wisdom and the modern world of anomic individualism that is dependent upon consumer-ism and the abstractions appearing on computer screens. The tensions between the time-tested forms of wisdom and the convenience and the immediate access to the data generated by experts whose long-range goal is to replace much of what is human with robots and machine forms of intelligence, is clearly articulated by a young woman who is herself caught between the two worlds. While pursuing a graduate degree at the University of Hawaii, she writes about how the cosmology that is the source of ecological wisdom is being threatened by the modern world's pursuit of progress. As she put it:

> He ali'i ka 'āina, he kauwā ke kanaka (the land is a chief, man is the servant) is a wise saying in our traditions. 'Āina encompasses the land, sky, ocean and all contained therein including plants, animals, and that which feeds and nourishes life. Our role as Kanaka 'Ōiwi ("Hawaiian") is found in genealogical relationship and responsibility to that which preceded us—plants/animals, our islands, the soils and waters that feed the plants/animals, to the eldest of the elements. Our responsibility extends to those who will come after us, our children and future generations. When our elders pass they are buried and they in turn become of the land. So we actively care for and protect our 'āina out of gratitude and survival because it feeds us physically; when we return to nurture and be nurtured by the 'āina it feeds us spiritually by restoring ancestral memory. Our ancestral wisdom/memory/traditions are alive today in our 'āina, in our elders, in our language, in our chants and song, in our na'au (where your navel is, your na'au are your guts, your soul).
>
> One of our beloved elders Pualani Kanahele reminds us all "I am this land, and this land is me." We perpetuate the love and respect our ancestors shared with their specific lands by telling their stories, continuing to grow native cultivars of taro, fighting for our waters and the inherent right to use them, through education—teaching our young, hands and feet in the soil and waters. We have a saying that is being used politically at the moment in protest to the building of telescopes on top our sacred mountains . . . "until the very last Aloha 'Āina" we will stand and ensure the continuity of our place and knowledge until the very last being that protects our land—through persistence, truth, and aloha.
>
> (personal communication, 9/1/15)

Where in the narratives of the computer scientists, data scientists, heads of corporations, agencies protecting the nation's security, and all the other individuals and groups who have now made data the highest form of knowledge, do we find any concerns about the lack of ecologically informed wisdom articulated in the above observations about what is being lost? The most abstract, that is, context-free, bits of information that are constructed on the basis of some expert's taken-for-granted cultural assumptions, who is often working for others higher up in the systems of economic and human exploitation, are supposed to guide decisions that will impact people's lives—people who are largely unaware of the shortcomings of data and the various ideologies that will guide its use. One of the great ironies of our times is that the traditions of ecologically informed wisdom are relegated to marginal status in our systems of higher education.

The other irony, for which everyone will pay dearly as the ecological systems begin to collapse, is that the possibility of finding from within our own western cultures the basis of an ecologically or even relationally informed wisdom tradition is being undermined by the values and knowledge given high status in our public schools and universities. As I observed in an earlier book, *The Culture of Denial* (1997), the high-status knowledge promoted in higher education is largely print-based and thus abstract, and increasingly computer mediated. It is also ideologically framed by the misconceptions of the 17th and 18th-century Enlightenment thinkers who promoted overturning traditions by relying upon critical thinking, scientific knowledge, and a secular worldview.

At the core of the high-status knowledge promoted in higher education are the deep cultural assumptions about the autonomous nature of the individual, a mechanistic and human-centered (anthropocentric) world, the progressive nature of change, the combination of cultural hubris and missionary spirit that justifies colonizing other cultures to adopt the core features of the western mindset that now accepts the replacement of humans, along with their traditions, with digital machines. Given these characteristics of high-status knowledge, and the increasing reliance upon computer-mediated thinking as a source of entertainment, there is little likelihood that either students or their professors will even be aware of the relational wisdom of Buddhism and Confucianism, and the ecologically informed wisdom traditions of indigenous cultures that are becoming increasingly aware of how western cultures are accelerating global changes that are threatening their future existence.

In spite of the continuing imprint of the anthropocentric message in the *Book of Genesis* on the consciousness of most Jews, Christians (especially fundamentalist Christians), and even the growing atheist movement that continues to adhere to more of the Judaic/Christian cosmology than they recognize, there are other obstacles to acquiring a shared wisdom tradition that would limit the excesses inherent in the libertarian/market liberal ideology that relies upon data to justify everyday decisions. Unlike many of the indigenous cultures that have developed ecologically informed wisdom traditions over the centuries, and embedded this wisdom in the many dimensions of their symbolic culture that guided daily practices, the West's guiding modernizing cosmology is interpreted from the perspectives of the many tribal traditions that have been melded into what is called western civilization—which is a high-status phrase that hides the tribal roots of various groups that occupied the territories we now call England, France, Italy, Poland, and so forth. Again, it's a matter of so-called autonomous individuals and their primary tribal roots being overwhelmed by the libertarian/market liberal myth of the role that data plays in achieving even more material progress.

There are writers such as Henry David Thoreau, Aldo Leopold, Rachel Carson, and Wendell Berry who provide key sensitivities and insights upon which wisdom traditions could be based. Whether youth will encounter their writing as they search the Internet or encounter them in the educational software written by the technologically minded programmers is problematic. And if they were to read any of them, each student would need to make her/

his own decision about taking them seriously when the consumer-oriented cultural ecology that impinges on their senses and behaviors communicates a different message: namely, that consumerism is still the main road to personal happiness and success. Data is the basis of this message, as well as the innovations that keep the economy expanding even as it shrinks the opportunities to work in a setting not dictated by the digital systems.

The final blow to a wisdom tradition becoming the primary moral guide that leads to reflecting on whether data-based decision making takes into account the primary responsibilities to the natural and cultural ecologies upon which we are dependent is that few people, even those who are highly educated, understand what is problematic about the origins and uses of data. It has now acquired a cult standing, which will only be strengthened as the digital revolution expands its influence over more aspects of daily life.

The final judgment is that the robust ecologically informed wisdom traditions that once guided how to live within the limits and possibilities of the local bioregions have now been largely overwhelmed. The central question today is whether the increasing emphasis on data and reliance upon new digital technologies will lead to an awareness that one of the central messages in the *Book of Genesis*, which is "Be fruitful, and multiply, and replenish the earth, and subdue it: and have **dominion** over the fish of the sea, and over the fowl of the air, and over everything that moveth upon the earth" (Genesis 1:28, italics added) is leading to the collapse of the ecosystems upon which we depend.

As computer scientists have announced that the transition to the age of singularity is now occurring, and that super-intelligent computers will take over as the world enters the post-biological phase of evolution, it will be up to computers to interpret what "dominion" means, and to find in their world of seemingly endless data the moral guidelines that will replace the ecological wisdom of the indigenous cultures that were long-term inhabitants of the land. Perhaps the next exodus should be from the Garden of Data and its tree of knowledge.

References

Bowers, C. A. 1997. *The Culture of Denial.* Albany, NY: State University of New York Press.
Lawlor, R. 1991. *Voices of the First Day: Awakening in the Aboriginal Dreamtime.* Rochester, VT. Inner Traditions International.
Smith, H. 1991. *The World's Religions.* New York: Harper One.
Vasquez, R. G. 1998. "The AYLLU." In *The Spirit of Regeneration: Andean Culture Confronting Western Notions of Development,* ed. Frédérique Apffel-Marglin. London: Zed Books.

16 Is the Digital Revolution Driven by an Ideology?

The answer to the chapter title is "Yes!" This question is likely to come as a surprise to the people who rely upon computers for reasons of personal convenience, such as myself, because of their usefulness in solving difficult problems, increasing productivity and profits, and carrying out other useful tasks. Current thinking among the general public and within the computer science community about the uses of digital technologies is limited largely to how they can be used more widely and effectively. If there is a consensus in this era of increasing armed conflicts, it is that digital technologies are the gateway to further progress and that the so-called digital divide between the users and non-users must bring the latter into the modern world.

If we are experiencing progress in achieving a better quality of life, it would seem pointless to muddy the waters by asking whether the digital revolution is driven by an ideology. Ideologies, such as the social justice liberalism which most democrats support, the market liberalism of the faux conservatives, fascism, and various religions that function in the same way as ideologies, share the same fatal flaw of not recognizing the world's diversity in cultural assumptions that frame how human with human and human with natural systems are to be understood and valued. That is, ideologies are colonizing conceptual and moral templates for how the world's people should think and behave toward others and the environment. Deviation from what is prescribed by the ideology or religion often leads to various forms of social, economic, and even military sanctions. Given the recent colonizing record of ideologies, how can the digital revolution, with its capacity of enabling people from diverse cultural backgrounds to solve local problems, to educate their children, and to become connected to the Internet and global markets, be understood as a cultural colonizing ideology that claims to embody universal truths?

To get beyond the current ideologically driven surface level of thinking that explains the digital revolution as having at last reached the exponential rate of development that matches Moore's Law, it is necessary to make an important detour into the realm of ecological linguistics. Indeed, if there is a universal process that challenges the certainties of an ideology, it is that all cultures are based on metaphorically based linguistic processes called mythopoetic narratives, religions, root metaphors, and symbolic constructions.

These metaphorical linguistic processes have a history rooted in diverse mythic stories of origins, powerful evocative experiences, and life-changing analogies where the past is carried forward in vocabularies that the present generation largely takes for granted. In cultures that value innovations that supposedly overturn the traditions of the older generation, the new ideas and innovations actually build upon and carry forward many of the deep cultural assumptions that underlie earlier patterns of thinking.

These patterns need to be understood as part of a culture's linguistic ecologies. Foundational to the West's linguistic ecology are the root metaphors of patriarchy, a human-centered world (or anthropocentrism), individualism, mechanism, progress, economism, and evolution. The root metaphor of ecology is now emerging as an explanatory framework that challenges many of the root metaphors that gave conceptual direction and moral legitimacy to the industrial culture that has now entered the digital phase of globalization. Like all ecological systems, these root metaphors have a history and now play a powerful role by introducing changes to the linguistic ecologies of other cultures. Most importantly, these root metaphors provide the current tacitly held interpretative frameworks for thinking about relationships. They also frame how to understand and solve problems—some of which might not exist if the culture relied upon other root metaphors. For example, the mythopoetic narrative (root metaphor) of the Quechua of Peru represents nature and humans as in a mutually nurturing relationship. Root metaphors in the West, such as "mechanism," are supported by other root metaphors such as "progress," "individualism," "an anthropocentric world," "economism" (profits), and "evolution." The root metaphor of "mechanism," which displaced the root metaphors of the feudal era, not only led to reframing how to think about government as based on systems of checks and balances, and the nature and functions of organs—such as the heart as a pump, how the brain operates—but has led to thinking of artificial intelligence as like human intelligence, and more recently to thinking of human intelligence as like computer intelligence. By relying upon the root metaphor of evolution, Ray Kurzweil, a leading computer scientist/futurist thinker (and proponent of Social Darwinism) is now claiming that humankind has now entered the post-biological phase of evolution, with super-intelligent computers taking over from humans as the world enters the singularity stage of natural selection.

What is especially important about root metaphors is that their supporting vocabularies exclude other vocabularies that would lead to different understandings of reality. The mechanistic root metaphor that now governs agriculture, education, and the organization of work, excludes the vocabularies necessary for giving expression to the sacred, the differences in cultural patterns of thinking and values, what is learned from reliance upon the senses, such human attributes as insights and empathy, and the tacit and taken-for-granted experiences that vary within different cultural contexts.

The root metaphor of mechanism, and the values that are consistent with this explanatory framework, are now leading to prioritizing efficiency and profits over the need of people to have access to employment. To cite another example, the root metaphor of "progress," which is supported by the vocabularies of other root metaphors such as "individualism" and "a human-centered world," exclude the vocabularies necessary for naming the traditions that need to be intergenerationally renewed as the twin crises of a rapidly degraded environment and the globalization of the digital revolution continue to contribute to the loss of ecologically sustainable forms of knowledge and skills. Many of these traditions enabled people to live less monetized lives, which also strengthened the patterns of mutual support that are the basis of the cultural commons of different cultures. This can be understood as an intergenerational gift economy which was and continues to be passed forward through face-to-face communication and mentoring. The initial misunderstanding that often occurs when first learning about the cultural commons is that it will involve returning to the lifestyle of earlier centuries. This is due to not recognizing that the cultural commons are different from environmental commons that were enclosed in early 19th-century English. The non-monetized intergenerational traditions, skills, and mentoring relationships can be traced back to the beginning of human history, and they continue to exist in every culture, community, family and human relationship—and even in the experiences of those who are committed to transforming what remains of the cultural commons into new market opportunities.

There is another common characteristic of all cultural and natural ecologies: namely, that there are no separate autonomous entities, ideas, things, facts, or individuals. Everything exists within complex webs of relationships and interdependencies. These relationships—whether at the micro and macro level—serve as the information pathways through which messages (which may be at the chemical, genetic, temperature, metaphorical, behavioral, and different semiotic pattern) are communicated. Within human cultures, the metaphorical nature of the culture's vocabulary influences whether the information being communicated through these relational pathways will be recognized—and how they will be interpreted. Metaphors, in short, can expand just as they can inhibit awareness and understanding.

For example, the root metaphor of progress, when combined with reliance upon the abstractions of the printed word rather than with what can be learned through the senses, now leads to ignoring the importance of many skills and mutually supportive relationships that were, for previous generations, common-sense understandings—or what can be referred to as tacit knowledge of shared cultural patterns. Other examples can easily be cited of how the metaphorical nature of language carries forward and reproduces the misconceptions and silences of earlier eras that limited understanding of interpersonal and environmental relationships. For example, the early analogs that framed the meaning of the metaphor "woman" reflected the prejudices of the era that excluded recognizing her as possessing the potential to be a painter, an historian, and generally highly intelligent and more physically fit in certain activities than men. What was being communicated in many male/female relationships was limited to what fit the conceptual framework, largely dictated by the prejudices encoded in the language that earlier generations took for granted.

The analogies that framed the meaning of other metaphors are now undergoing change, just as we are starting down the pathway to understanding intelligence as relational and ecological. Less understood is that words have a history. When their meaning is framed by the analogs settled upon in the past, they carry forward earlier forms of cultural intelligence as well as the era's misconceptions. The choice of analogs that are ecologically informed about environmental issues and an understanding of the cultural commons can lead, for example, to changing the meaning of wealth from that of possessing money to that of possessing useful skills and patterns of mutual support that strengthen community.

That print is a technology that provides only a surface knowledge of a world that it represents as static rather than as emergent is yet another challenge, given the long history of associating print with literacy, democracy, and becoming civilized. The challenge of recognizing the conceptual framework promoted by print is made more difficult by the way in which print reinforces the abstract theories of Western philosophers and social theorists as based on a rational process that supposedly is free of hidden cultural assumptions. Print-based abstract thinking avoids the complexities of the senses, communal memory, differences in cultural and natural contexts, and the questions that arise when it is acknowledged that the language systems that are the basis of print-encoded cultural storage and communication are based on root metaphors and mythopoetic narratives. That most writers, computer programmers, and ideologues are unaware of the need to make explicit the tacit patterns of thinking and deep assumptions of their own culture too often results in the printed word representing only a partial understanding. This leads to the all too real habit of assuming what appears in print to be an objective and factual account rather than being the writer's interpretation that, in turn, was influenced by the linguistic ecology that was the basis of her/his socialization.

For many Western readers, there is another linguistic convention that is likely to influence whether any of this will be taken seriously: namely, the either/or convention of thinking that excludes the possibility that ideas, technologies, policies, and so forth, may exhibit short-term gains while also leading to destructive consequences at a future time. For example, many new digital technologies represent short-term gains in empowerment and achieving greater efficiencies. However, when considered within a larger context, such as how they contribute to the ecological crisis and to increasing levels of poverty and unemployment among the world's population now moving toward the 9 billion mark, their benefits must be weighed against the loss of important forms of knowledge, skills, and social justice traditions. And when we reach the critical point where current social systems are no longer able to cope with the dimensions of the crisis, moral judgments that are beyond the capacity of computer systems will be needed.

What is ironic is that two of the most prominent features of all cultures— the use of a metaphorical language rooted in the symbolic history of the culture and reliance upon technologies—are not required areas of study for all students. Thus, the students' current lives and future prospects are being rapidly reduced by cultural forces of which they have little understanding. Yet, their formal education leaves them with the mythic understandings formed during the last 500 or so years when the forces of industrialization began the shift to an individualistic, mechanistic, and consumer-dependent lifestyle. Unfortunately, their classroom teachers and professors, who continue to reproduce the interpretative frameworks of their own mentors who under- stood environmental issues as the responsibility of scientists and technologists— and thus free of cultural influences—failed to provide the educational basis for recognizing that the modern mythic understanding of language as a conduit in a sender/receiver process of communication, and technology as culturally neutral, are not ecologically sustainable.

When we begin to recognize how patterns of thinking and behavior always exist in a relational world, and that the relationships serve as complex information pathways crucial to whether the relationships lead to destruc- tive outcomes (like an ecology of weeds, as Gregory Bateson put it) or con- tribute to enhancing the life-forming and sustaining processes of the Other, it then becomes possible to take seriously the question about whether the digital revolution is driven by an ideology. This question leads to an even more important question: namely, is this ideology based on cultural myths that undermine the ecologically sustainable forms of knowledge and values of other cultures that have taken a different and, in many instances, a more ecologically informed, approach to development?

Leading computer scientists and futurist thinkers exhibit absolute certainty about the nature of the forces driving the digital revolution. But they do not recognize these forces as expressions of an ideology. For them, science pro- vides the best explanation of why these forces are irreversible as well as why they should be understood as dictating the fate of all cultures. When they are speaking in the language of science, they call this force evolution. And when writing about its impact on other cultures, they revert to the high- status vocabulary of their Western culture by referring to these forces as the expression of progress. They exhibit little awareness that they are moving down the slippery slope of scientism. Thus, what they refer to as Nature's process of evolution that dictates that computer intelligence has entered the era of singularity, where human intelligence is being surpassed by computer intelligence, turns out to be the Social Darwinism that has played such an important ideological role in the winner-take-all mentality that dominated past and current periods in American capitalism.

It is important to recognize how they adapt Darwin's theory of natural selection as a way of explaining why the digital revolution is leading to the extinction of the world's diversity of cultures, which are to be replaced by the emergence of a hybrid where super-powerful computers rely upon Western assumptions to collect and process data, identify and solve problems, and generally replace the culturally diverse cognitive and moral abilities of humans. That these computer/futuristic thinkers understand evolution as leading to the elimination of the world's diversity of cultures and to replacing them with the monolithic nature of computer intelligence that is unable to encode and process the tacit, contextual, and taken-for-granted patterns of different cultures—including the patterns of moral reciprocity, empathy, intersubjective identities, and wisdom traditions—should be one of the warning signs that these leading computer scientists/futurist thinkers do not understand one of the most widely recognized characteristics of evolution. When the explanatory power of evolution is not based on the Western assumption of a linear form of progress, it then provides a way of understanding that nature depends upon diversity in determining what represents the better-adapted genes and behavioral traits.

Another major source of confusion shared by the computer scientist/futurist thinkers is that they do not understand that memes do not have the same scientific basis that genes have in the scientific world. They simply accepted a metaphorical slight-of-hand word trick initiated by Richard Dawkins, and supported by E. O. Wilson and other prominent Social Darwinian thinkers who argue that memes play the same role in the evolution of cultures as genes play in the biological world. The problem is they do not recognize that the use of Social Darwinism, as a conceptual framework for deciding what represents backward and thus less evolved cultures that can be replaced by the globalization of computer-mediated intelligence, is also based on the cultural assumption about the progressive nature of change. For these futurist thinkers, it is assumed that the corporations that rely upon big data, the connected world of the Internet, and technologies that track people's behaviors are more evolved than indigenous cultures that developed place-appropriate technologies, the arts essential to communicating about the reciprocal nature of relationships, patterns of mutual support, and an ecologically informed spirituality that enables them to live within the limits and possibilities of the bioregion. As we shall see, leading computer scientists/futurist thinkers refer to these cultures as backward and moving toward extinction while viewing intelligent self-programming machines, and the virtual worlds they can create, as being carried forward by the process of evolution.

In the mid-1980s, when digital technologies were just being promoted by computer scientists and the Willy Lomans of the computer industry as essential to students constructing their own knowledge and staying connected with others, Hans Moravec wrote *Mind Children: The Future of Robots and Human Intelligence* (1990). This book was intended as a wake-up call about how evolution dictates that computers, including robots, were on the verge of replacing humans, with all their physical limitations and inefficiencies. By explaining that the coming extinction of all humans is dictated by Nature's agenda for ensuring that the better adapted survive, it gave Moravec's statements the appearance of a high degree of scientific legitimacy.

At that time, few members of the public were aware that Moravec was helping to lay the conceptual and moral foundations for the introduction of digital technologies created by computer scientists and engineers who were and continue to be largely indifferent to the unintended cultural consequences of their inventions. Indeed, according to the Social Darwinian conceptual framework upon which Moravec relied, the introduction of robots and other digital technologies that reduce the need for workers, and thus their ability to practice a craft and to earn a living, is dictated by Nature's logic. This same logic, along with the capitalist's greed for increasing profits, also dictates eliminating the benefits and social contracts won in earlier labor struggles.

Other losses that Moravec viewed as a necessary consequence of computers replacing humans in the process of evolution include the forms of intergenerational knowledge and achievements essential to civil societies that have learned to live by social justice principles. Privacy, intergenerational knowledge, skills, and mentoring relationships essential to mutually supporting communities are also part of the taken-for-granted traditions of some cultures. Unfortunately, these tacit and contextually based aspects of culture are not what self-programming computers and robots are particularly good at replicating. Survival of the fittest, the phrase coined by Herbert Spencer to explain a key feature of Darwin's theory, dictates that super-intelligent computers are to replace humans with all their vulnerabilities. Spencer's phrase has now been replaced by the less ruthless-sounding phrase of "better adapted," as it is more easily accepted by the public conditioned to equate improvement with progress. And who can be against progress?

Ray Kurzweil, perhaps the most widely recognized and acclaimed computer scientist/futurist thinker, published *The Age of Spiritual Machines* (1999). This book assured, again with the certainty that has become a hallmark of this genre of thinkers, that computers would evolve to the point where they will replicate all aspects of human experience, including having religious experiences. Thus, there would be no reason for humans to become anxious about their coming extinction—which after the final transition would become a non-issue.

Gregory Stock, whose degree is in the field of biophysics, was one of the earliest to predict, to use the subtitle of his book, "the Merging of Humans and Machines into a Global Superorganism" he named "Metaman" (1993). The diversity of the world's cultures, as he put it, "is mostly a thing of the past." The archaic forms of knowledge of these non-Western cultures are being replaced by the evolution of Metaman's ability to 'think' by using a 'brain' that is literally all around us. And that brain contains within it the functional equivalent of a global 'memory' housing all of humanity's accumulated knowledge. Examining the evolution of this global memory, he concludes, "reveals its nature and future" (p. 85).

It is important to recognize that Stock's Social Darwinism locates the forces of change outside the realm of human decision-making. By extending Darwin's theory to include the evolution of cultures, the diverse cultures of the world will have no role in deciding if they are willing to be part of the great extinction that will follow the further evolution of digital technologies. Cultures headed for extinction, like the emerging Metaman, must simply accept what the process of natural selection dictates. And dictate it will! As Stock describes this transformation to human–machine hybrids: "as the nature of human beings change, so too will the concept of what it means to be human. One day humans will be composite beings: part biological, part mechanical, part electronic" (p. 152).

It is important to mention Ray Kurzweil again, as he has received a number of honorary doctorates, national awards, and large sums of money for his digital inventions. He is clearly a highly inventive computer scientist, and is equally acknowledged as a leading futurist thinker. His book, *The Singularity is Near: When Humans Transcend Biology* (2005), not only gives an account of the stages in which humans will be replaced in the process of evolution, but also the approximate dates. That the influence of the theory of singularity that represents a fundamental transition in the evolutionary process from a human/biological world to that of digital machines is being taken seriously by other computer scientists can be seen in the number of young computer scientists who enroll at Singularity University, which is located on a campus near Google, where Kurzweil is one of the leading engineers and innovators.

Kurzweil's reliance upon Darwin's theory of evolution is clearly evident in his predictions about the cognitive take-over by digital technologies. The following represent just four of a long list of changes that will be brought about as we enter the era of singularity.

> With both hardware and software needed to fully emulate human intelligence, we can expect computers to pass the Turing test, indicating intelligence indistinguishable from that of human intelligence, by the end of 2020.
>
> When they achieve this level of development, computers will be able to combine the traditional strengths of human intelligence with the strengths of machine intelligence. . . .
>
> Machine intelligence will have complete freedom of design and architecture (that is, they won't be constrained by biological limitations, such as the slow switching speed of our interneural connections or a fixed skull size) as well as consistent performance at all times.
>
> (Kurzweil, 2005, pp. 25–26)

In Kurzweil's version of digital heaven, which he calls virtual reality, we will be able to enter and explore realities that are radically different from the world of culturally embodied experiences. As we enter these virtual realities, "we won't be restricted to a single personality, since we will be able to change our appearance and effectively become other people . . . We can select different bodies at the same time for different people. Your parents may see you as one person, while your girlfriend will experience you as another" (2005, p. 314). While this last projection of life in virtual reality does not seem much different from what many parents now experience, it is important to recognize that Kurzweil ignores cultural differences, which involve differences in belief and values systems—as well as differences in personalities within these different cultures.

A further example of Kurzweil's reductionist thinking (or what can be referred to as his abstract representation of human intelligence as though it is universally the same for all people) can be seen in the title of his 2012 book, *How to Create a Mind: The Secret of Human Thought Revealed.* One does not have to read beyond the title of the book to recognize the dangerous combination of a person who is promoting a fundamental change in the world's cultures and whose thinking is based on the misconception that there is only one form of human intelligence. This is a chief characteristic of an ideologue.

In the epilogue to *How to Create a Mind*, Kurzweil writes that "the last invention that biological evolution needed to make—the neocortex—is inevitably leading to the last invention that humanity needs to make—truly intelligent machines—the design of one inspiring the design of the other" (2012, p. 281). Given that there are many forms of human/cultural intelligence, and given the hundreds of languages still spoken in the world, the question then becomes: which form of cultural intelligence will inspire the design of machine intelligence? Will it be fundamentalist Christian, Muslim, Hopi, Buddhist, market liberal?

The title of other books by computer scientist/futurist thinkers also reveals the same assumption that the globalization of the digital culture will lead to progress for the entire world. These include *Darwin among the Machines: The Evolution of Global Intelligence*, by George Dyson (1998); *Abundance: The Future is Better than You Think*, by Peter Diamandis and Steven Kotler (2012); *The New Digital Age: Reshaping the Future of People, Nations, and Business*, by Eric Schmidt and Jared Cohen (2013); *Radical Abundance: How the Revolution in Nonotechnologies will Change Civilization*, by K. Eric Drexler (2013); *Facing the Intellectual Explosion*, by Luke Muehlhauser (2013); and *The Second Machine Age: Work, Progress, and Prosperity in an Time of Brilliant Technologies*, by Erik Brynjolfsson and Andrew McAfee (2014).

Among this latter group, only Dyson unequivocally embraces the Social Darwinian conceptual framework that is such a prominent part of the thinking of Moravec, Stock, and Kurzweil. The others, while referring to evolution, rely more on the Western root metaphor that equates technological innovations with the market liberal way of understanding progress. The computer/futurist thinker's way of understanding progress is not like that of the Western Apache who interpret progress as achieving wisdom by avoiding the distractions of the personal ego and the demands of the external surroundings—including the expectations of others, or that of a Buddhist in attaining a mindful existence, or that of other non-Western cultures less focused upon turning all aspects of daily life into expanding markets and profits.

The market liberal ideology of these computer/futurist thinkers aligns perfectly with Social Darwinian thinking, as they both are dependent upon other cultural assumptions (root metaphors) such as the autonomous nature of the individual, a human-centered (which is to become a computer-centered) world, mechanism, and economism (which holds that everything has an economic value). Other assumptions that support the market liberal ideology include accepting a sender/receiver view of language that hides that words have a history and carry forward the cultural misconceptions of earlier eras (which supports the myth of objective data, information, and ideas such as free markets and private property), and the progressive nature of conflict and competition in overcoming what is regarded as inefficient and tradition-bound. These digital revolutionary ideologues also exhibit what Wendell Berry termed the growing imperialistic agenda of science, and thus the silences, prejudices, and reductionist thinking that accompanies how culture is understood by mainstream Western scientists.

Implications of the Ideology that Drives the Digital Revolution

One of today's ironies is that the computer/futurist thinkers totally ignore what the environmental scientists (who rely upon increasingly sophisticated digital technologies) are reporting; namely, that there is a rapidly deepening ecological crisis. Their silence aligns their thinking with that of the corporate and think tank market liberals who are in denial that the industrial/consumer being globalized is undermining the self-renewing capacity of natural systems. They also share the limitations of the scientists' way of understanding the ecological crisis, which is to promote the development of new less environmentally disrupting technologies. What is being ignored are the cultural roots of the ecological crisis, as well as an understanding of cultures that have taken more ecologically sustainable paths to development. These are the cultures that are to disappear when the era of super computers and global connectedness (singularity) takes over.

Ideologies, as mentioned earlier, are sustained by supporting root metaphors and vocabularies, which also serve to exclude other vocabularies and interpretative frameworks. The narratives supporting the ideology of the digital revolution will be heard differently within different sectors of society—with the medical, industrial, agricultural, military, and educational sectors learning to expect further innovations ahead that will increase their efficiency, problem-solving abilities, and profits. What these ideologically influenced narratives will not address are the cultural traditions that have a smaller ecologically destructive footprint and thus should be intergenerationally renewed.

There is a reason for the silence on the part of the computer/futurists that goes beyond the myopia and hubris of their ideology. That is, if their educational backgrounds were to be studied, it is highly likely that we would find that they did not learn about the tacit interpersonal norms and conceptual/linguistic patterns of their own culture—and the many symbolic ecologies that sustain and transform the diverse cultural traditions. Such a study would reveal that they, as a group, think of traditions in the most simplistic and reductionist terms—even though their everyday lives involve the unconscious reenactment of cultural patterns that can be called traditions. Similarly, a study of the educational background of most scientists would reveal the same lack of knowledge of whether the cultural ecologies they participate in on a daily basis contribute to a sustainable or unsustainable future.

There is now a growing understanding that the world, from the micro to macro levels of natural and cultural ecologies, is one of emerging relationships that serve as multiple pathways of information exchange. Digital technologies, aside from being limited by the cultural patterns that are not made explicit, cannot represent the emergent world of relationships—except at an abstract level where differences in ways of knowing are ignored. Even what is streamed is an event taken out of the context of the cultural ecology that has a history of interactive influences—including the cultural assumptions that frame how the differences which make a difference in relationships are interpreted. When both natural and cultural ecologies are understood as emerging relationships, the ability to recognize what is being communicated through these relationships becomes more critically important. This also requires recognizing how the vocabularies inherited from the past may limit awareness of the information being communicated within and between cultural and natural ecological systems.

A more immediate set of issues that the digital ideologues are ignoring can be traced to their indifference to the changes they are introducing into other cultures. One of the consequences of their formulaic thinking, which leads to equating new digital technologies with progress, results in their not considering the importance of the cultural traditions that are being lost. The push to develop smart technologies that will enable governments to control the flow of traffic, as well as enable the police to engage in real-time law enforcement, is just the start of the computer industry's effort to introduce sensors into all built environments for the purpose of collecting data on every aspect of human behavior. Just as Jacques Ellul predicted in his 1964 classic, *The Technological Society*, technological progress in the West will move from helping to solve problems, including crimes, to anticipating and instituting ways of controlling how they will occur in the future. This shift on the part of technocrats from responding to the diversity of people's culturally influenced behaviors to creating digital systems that limit their behaviors in ways that fit criteria that have not been determined by the democratic process, but instead by the ideology that interprets how the data is to be used to create more efficient systems of control, can be seen in the recent efforts of European officials to require all imported cars to feature a built-in mechanism that will enable the police to stop vehicles remotely. This approach to progress leads in turn to asking whether collecting data from sensors that keep behaviors in every part of the household under constant surveillance, which will be justified on the grounds that the data will lead to people's ability to make healthier and safer decisions, will be used by corporations to promote their life-enhancing products. That is, is there an economic interest that promotes these total surveillance systems, or is it the further reach of the National Security Agency?

In *To Save Everything, Click Here: Technology, Solutionism, and the Urge to Fix Problems that Don't Exist*, Evgeny Morozov (2013) cites several examples of technological progress based on the assumption that anticipating and correcting the future misbehavior of people must be build into the technology. Apple, for example, recently patented technology that deploys sensors inside the smartphone that measure whether the car is moving, and if the person is both driving and using the phone, the phone's texting capabilities will be blocked. Another example of bringing behavior in line with the norms of the people who create the technology is the Project Mobil system being created by Intel and Ford. It involves a face recognition system that will prevent the car from starting and will send the picture to the car owner if the system does not recognize the face of the person turning on the ignition system.

The same drive to use massive amounts of data, and the connectivity between digital systems, can be seen in the current effort to reduce the depth of knowledge that students should be learning to the supposedly objective bits of information and facts that can be machine scored. The data from this reductive process can then be used to determine the teacher's effectiveness in raising test scores. The cultural issues that cannot be reduced to measurable data, including the diverse ethnic and economic backgrounds of both the teacher and students, is simply ignored when in reality it may have the greatest influence on student learning.

The massive amounts of data now collected as part of the national effort to identify potential terrorists, even when the definition of who is a terrorist is open to ideologically driven interpretation, is already limiting the expression of ideas critical of the excesses of corporate America. As data is only a surface and fragmentary representation of the cultural context from which it is taken, it is also open to being interpreted differently—depending on the interpretor's ideology. And if the ideology is based on assumptions that experts have the best answers, and on abstract assumptions about the progressive nature of technologically driven and monitored change, then the next step is to incorporate the principles of behavior modification. That is, the system provides data on a person's behavior, shows how it compares with the performance of others, and provides the winner with a tangible reward. The use of behavior modification techniques represents a top-down system of control that is justified by the experts in the name of progress.

As briefly noted above, the modernizing ideologies driving the digital revolution continue to carry forward the Enlightenment misunderstandings about the nature of cultural traditions, particularly those traditions of the cultural commons that enabled people to live more community- rather than individually-centered lives. The importance of the cultural commons that continue today to be passed forward through face-to-face communication and through mentoring relationships—which encompass the culturally diverse approaches to food, ceremonies, creative arts and craft knowledge, knowledge of the life cycles in the local bioregion, the traditions of civil liberties slowly gained and easily lost, and even language itself—enable people to live less consumer-dependent and thus less ecologically destructive lives.

The problem with the Enlightenment thinkers who helped to put the West on the pathway to integrating science, technological innovation, and the industrial/market system of production, is that they were unaware of environmental limits and thus were unable to recognize what most computer scientists still do not recognize. Namely, that the further enclosure of what remains of the cultural commons by digital technologies, especially in an era when progress is understood as the further computerizing of the workplace, will lead to more poverty and eventually social unrest that will require the emergence of a Stasi-style police state where everybody is being watched. The data that provides the government all it needs to know about people's lives, as well as the tracking technologies and military-style hardware used by law enforcement, are already in place.

The deepening ecological crisis—which is leading to shortages of water for agriculture and even for meeting basic human needs, the extreme changes in weather that are devastating lives, and the changes in the chemistry and temperature of the world's oceans (along with over-fishing) that are reducing people's access to protein—is accelerating. The deep cultural assumptions that gave conceptual direction and moral legitimacy to the first Industrial Revolution, and now to the second digitally driven revolution, should be the focus of educational reform, especially at the university level. These assumptions originated in the abstract thinking of Western philosophers and social theorists, and achieved a status that placed them beyond questioning as scientists, technologists, and capitalists relied upon these assumptions in creating new sources of wealth, personal conveniences, and higher standards of health and longevity. These assumptions were based on the idea that natural resources are unlimited, and if limitations do occur, scientists will be able to create alternatives. In the West, the dominant idea was that progress would not be limited as long as scientists, technologists, and capitalists were freed from the traditions of the past. The problem with this way of thinking is that it led to an indifference to understanding how to make the transition to an ecologically informed form of consciousness, even though the market system and the new digital technologies contribute to changing consciousness in ways that are even less ecologically sustainable.

Basically, the digital revolution perpetuates the limitations found in the thinking of the Western philosophers and social theorists who provided the original conceptual and moral scaffolding that supports what has now become a global economic and technological agenda. These philosophers and social theorists relied upon print to communicate their ideas, and in the process ignored the fact that the meanings of words they used were framed by the analogs settled upon in the past, and that the new meanings, such as how to understand "free markets," the nature of "property," "data," "woman," and so forth, were the outcome of debates about what constituted analogs that met the criteria of progressive and scientifically informed thinking. To reiterate a point made earlier, the cultural emphasis on literacy, and now on the empowering nature of print that is read on the computer screen (which does not substantially differ from the print appearing on paper), hides what is now needed as the ecological crisis deepens. Namely, how to recognize that the meanings of most words appearing in print (as well as spoken) have a history, and that they encode many of the misconceptions and silences of the earlier eras when the analogs were settled upon. These misconceptions and silences included the failure to recognize environmental limits, other cultural ways of knowing (including cultures that had developed different forms of ecological intelligence), and thus the necessity of making the cultural turn away from the current consumer-dependent lifestyle—and toward a more cultural commons-centered lifestyle.

The printed words appearing on digital screens, which reinforce the mistaken idea that communication is like a sender/receiver conduit through which ideas and "objective" data and information are passed, also marginalize awareness of the metaphorical nature of language—including how earlier misconceptions become the basis of the individual's supposedly autonomous thinking. As more of the formal educational processes are mediated by cultural amplification and reduction characteristics of digital technologies, as well as being mediated by the mindset of the people whose thought processes appear on the screen as objective and factual, there are few professors and even fewer classroom teachers who can explain to students how print, including English nouns, are unable to represent the emergent, relational, and co-dependent nature of the cultural and natural ecologies that make up their world. Both print and English nouns are unable to represent the full and emergent nature of living contexts where there are no isolated events, ideas, things, or data. Everything, when understood within an ecological framework, is emergent and responsive to the information/semiotic-rich exchanges occurring in living systems. Gregory Bateson's reference to the "differences which make a difference" is another way of understanding that life-sustaining processes, that is, behaviors, introduce difference to which the Other responds, and the response of the Other introduces differences that, like a wave moving across a pond, serve as sources of information that bring about changes in the entire cultural ecology.

And how many professors who find the digital technologies highly useful in their research, and in communicating with colleagues on a world-wide basis, can explain the metaphorical nature of language, and how different cultures are based on different root metaphors and mythopoetic narratives that in many cases lead to valuing oral forms of renewing intergenerational knowledge and skills? Many of these cultures live on the margins where there is no room for experimenting with new ideas and technologies. The allure of becoming modern by relying upon technologies leads the youth in many of these non-Western cultures to reject the knowledge and skills of the older generation as sources of backwardness. Unfortunately, they do not realize that the unemployment levels among their peers, which reaches 40 percent and above, may have something to do with becoming dependent upon a money economy and an increasingly computerized industrial system that can produce massive amounts of consumer goods. Traditions, which progressive ideologues dismiss as obstacles to achieving a better future, are for these cultures the basis of a subsistence existence, and a cultural commons that may be rich in the arts and patterns of mutual support.

The loss of privacy, craft knowledge, and employment opportunities as more forms of work are computerized, historical memory and awareness of traditional patterns of mutual support, and even the awareness that people possess levels of self-reliance and good judgment that preclude the necessity of being under constant surveillance, suggest that the progress-oriented ideology of computer scientists has its roots in their failure to understand the cultures into which their technologies are being introduced. If they understood a basic characteristic of cultural traditions, they would then possibly be aware of the fact that when a tradition is overturned as a result of a technological innovatiom, it cannot be recovered. For example, the tradition of privacy, which has now been lost to the progressive thinking of computer scientists, cannot be recovered. Nor can the personal confidence previously associated with engaging in private economic transactions be recovered now that the digital technologies enable hackers and the unemployed sitting in Internet cafes to steal the identity and resources of others. And who is going to take responsibility when cyber attacks disrupt the financial, energy, transportation, and other critical infrastructures? This problem cannot simply be dismissed by claiming that progress always involves unintended consequences.

There have been many genuine gains from the digital revolution, but at the same time it is leading both to new forms of personal fear and insecurity, and to an inability to recognize one of the realities that the ecological crisis will force everyone to recognize if we are to avoid the endgame of social chaos as systems begin to fail. The reality is that we need to begin thinking about conserving habitats, species, the forms of the cultural commons that reduce dependency upon consumerism, and the levels of toxins it introduces into the environment. And this imperative leads back to the current failure of computer scientists and programmers to understand how language both illuminates possibilities while hiding others. Whether it is the vocabulary of libertarianism and market liberalism or the vocabulary of conservatives in the traditions of Edmund Burke or the environmental/cultural conservatives in the tradition of Wendell Berry and Vandana Shiva, mindfulness and thus caution is required when considering the long-term implications of new (especially abstract) ideas and innovations—especially when the innovations lead to economic advantages for the groups hiding behind the rhetoric of progress and other god-words. So far, the dominant ideology driving the digital revolution only illuminates the short-term gains and leads to ignoring what needs to be conserved if there is to be a future for humankind.

These cautionary observations are not likely to be taken seriously by computer scientists, venture capitalists, and corporate CEOs constantly in search of new market opportunities. Many of these progress-oriented thinkers have made vast fortunes from the combination of commercial hype, blind faith in the myths of market liberalism, and the creation of digital technologies that are embraced by special interest groups seeking more effective ways to achieve their agendas—including corporations and the surveillance agencies of government. The public's addiction to being connected to the Internet has also contributed to their fortunes. The computer scientists' libertarian and market liberal way of thinking about the causes of poverty may also figure into why computer scientists are working to computerize as many skills and cognitive functions as possible. With a recent prediction that 47 percent of jobs in the West may be replaced by digital technologies within the next two decades, it would seem that there should be a debate among computer scientists about their contribution to world poverty and the growing social unrest.

Similarly, with the increasing public concern about the development of digital technologies that collect data on nearly every aspect of human behavior, and with this data stored by governmental agencies and used by corporations to promote their products, it would seem that computer scientists should begin to ask questions about whether there are moral and political guidelines that should limit their research and development. Their collaboration with the pharmaceutical industry and with scientists in the field of brain research, where one of the primary goals is to develop digital technologies that will bring more aspects of the individual's thought and behavior under external control, should also prompt an ongoing debate about the moral and political responsibilities of computer scientists. Some of these new technologies lead to genuine benefits. But many, such as the efforts to anticipate the thoughts and behaviors of others, are genuine threats to our traditions of civil liberties. The computer scientist/futurists who are the most dogmatic Social Darwinian thinkers are totally silent on this issue.

The writings of the computer/futurist thinkers, as well as the promoters of providing a computer for every child in the world, reveal a total lack of awareness of how digital technologies undermine the face-to-face, orally communicated symbolic traditions of non-Western cultures. As pointed out above, the virtual world of the Internet is also a world of abstractions that only connect in highly selected ways to those aspects of everyday life that have been made explicit, and experienced from the limited perspective of an expert whose real agenda is not always known. The complexity of information communicated through embodied relationships, which range from being able to respond appropriately to the cultural norms governing the tacit patterns of footing and framing that occur in all interpersonal relationships to the wealth of intergenerational knowledge that sustains the non-monetized traditions of the cultural commons, are now seen by many older members of these largely orally based cultures as being subverted as their youth become more dependent upon digital technologies. In effect, the globalization of digital technologies and the market system of production and consumption is being seen as a form of cultural and economic colonization by many adults who still possess a memory of their pre-digitized past. Just as many in our society would engage in armed resistance if Sharia law and a tribal system of government were imposed on our country, it should not be surprising that the modernizing agenda of the digital/market ideologues is also being resisted. Defeating the armed resistance to the West's colonizing agenda actually serves to increase the profits of our defense industry and the computer scientists who are now an indispensable part of this industry.

Summary

The issues raised here should be part of a national conversation—indeed, an international conversation. Given how the current system of Western education continues to privilege the patterns of thinking and values that perpetuate the now digitally driven industrial/consumer-dependent culture that is increasing unemployment, real poverty associated with the lack of protein as well as the poverty that accompanies the loss of the local cultural commons, and changing the chemistry of natural systems, it is hoped that this conversation will be given more than token recognition. The real hope is in the move toward local community-centered approaches to growing food, becoming energy independent, practicing local democracy, revitalizing the cultural commons that also include the hard-won traditions of civil liberties and social justice achievements, and the moral language governing relationships within the local cultural ecologies and those of the larger natural world. The focus would then shift from assuming that technological and profit-driven progress is the way forward to recognizing that we need to make conserving the intergenerational traditions that are ecologically sustainable integral to how we understand progress—which ultimately cannot be separated from an ecologically sustainable future.

References

Bateson, G. 1972. *Steps to an Ecology of Mind*. New York: Ballantine Books.

Brynjolfsson, E. and McAfee, A. 2014. *The Second Machine Age: Work, Progress, and Prosperity in a Time of Brilliant Technologies*. New York: W.W. Norton.

Diamandis, P. and Kotler, S. 2012. *Abundance: The Future is Better Than You Think*. New York: Free Press.

Drexler, K. 2013. *Radical Abundance: How a Revolution in Nanotechnology Will Change Civilization*: Nook e-book.

Dyson, G. 1998. *Darwin Among the Machines: The Evolution of Global Intelligence.* New York: Basic Books.

Kurzweil, R. 1999. *The Age of Spiritual Machines: When Computers Exceed Human Intelligence.* New York: Viking.

_____. 2005. *The Singularity Is Near: When Humans Transcend Biology.* New York: Viking.

_____. 2012. *How to Create a Mind: The Secret of Human Thought Revealed.* New York: Viking.

Moravec, H. 1990. *Mind Children: The Future of Robot and Human Intelligence.* Cambridge, MA: Harvard University Press.

Morozov, E. 2013. *To Save Everything, Click Here: Technology, Solutionism, and the Urge to Fix Problems that Don't Exist.* New York: Public Affairs.

Muehlhauser, L. 2013. *Facing the Intellectual Explosion.* Kindle e-book.

Schmidt, E. and Cohen, J. 2013. *The New Digital Age: Reshaping the Future of People, Nations and Business.* New York: Alfred A. Knopf.

Stock, G. 1993. *Metaman: The Merging of Humans and Machines into a Global Superorganism.* New York: Doubleday.

Wilson, E. 1998. *Consilience: The Unity of Knowledge.* New York: Alfred A. Knopf. OR.: Eco-Justice Press.

_____. 2014. *The False Promises of the Digital Revolution: How Computers Transform Education, Work, and International Development in Ways that Undermine an Ecologically Sustainable Future.* New York: Peter Lang.

17 Is the Digital Revolution Sowing the Seeds of a Techno-Fascist Future?

Before criticizing the title of this chapter as excessively alarmist, compare the guiding ideology of the digital revolution, as well as the cultural changes it has already introduced, to the characteristics of fascism. It is also important to recognize that fascism varies between cultures. Italian fascism was different in several important ways from German fascism, and if Oswald Moseley had come to power in Great Britain, his brand of hyper-nationalism would have differed—just as the fascism of France's Jean-Marie Le Pen, as it evolves, will be imprinted with what is distinctively French. The same holds for the early signs of techno-fascism, the chief characteristics of which suggest a more international and thus less distinctly American model.

The connections between technologies (digital, especially) and fascism are not widely recognized, partly because fascism is mostly understood by looking through the rearview mirror of recent historical events. Technologies were essential to the short-lived successes of both Italian and German fascism, but both were also driven by the social unrest following the end of World War I, racial mythologies (especially for the Germans), the lack of well-established democratic institutions, and the economic turmoil of worldwide depression. Techno-fascism is characterized by the way more aspects of daily life are becoming dependent upon digital technologies. This leads to many benefits, but at the same time, digital technologies not only reduce the diversity in cultural ways of knowing; they also increasingly subordinate human thought and behaviors to the dictates of machines. Unlike the racist mythologies of German fascism, the mythic dimensions of techno-fascism are rooted in ancient religious narratives about humans naming and taking control of the environment, and in the abstract thinking of philosophers who laid the conceptual and moral foundations for the modern myth of progress—which includes the idea that human life is mechanistic and driven by Nature's laws governing natural selection. While the moral foundations of techno-fascism align with the values of market capitalism and the progress-oriented ideology of a science that easily slips into scientism, its level of efficiency and totalitarian potential can easily lead to repressive systems that will not tolerate dissent—especially on the part of those challenging how the colonizing nature of techno-fascism is destroying the environment and alternative cultural lifestyles.

The primary characteristic of all fascist modernizing movements is conformity of thinking and behavior, which is directed and controlled by total surveillance systems that track and keep records of people's thoughts, behaviors, and relationships. The latest in the emerging techno-fascist arsenal of surveillance technologies is the facial recognition system being adopted by local police—which will shortly become part of the FBI's one billion dollar Next Generation Identification Program. Thus, photos of people not suspected of criminal activities, as well as those who are, will be instantly available to 18,000 local, state, federal, and international law enforcement agencies. Facial recognition technology can identify 16,000 distinct features of a person's face and compare them with other photos held by police agencies at a rate of more than one million faces per second.

Three of the most important threats to what remains of our civil liberties include: (i) facial recognition software has a 20 percent failure rate (ii) social unrest resulting from extreme environmental changes can easily lead to redefining what constitutes criminal behavior, and (iii) police biases and misinterpretations lead to police actions that result in the deaths of innocent people (a problem now plaguing local police across the country).

In spite of how human biases influence the interpretation of data and images, a characteristic of fascism shared with the market system is the quest for new technologies that make tracking people and their market preferences more efficient. Writing in the *New York Times* (9/25/15), John Eligon and Timothy Williams note that corporations are also turning to what in policing circles is called predictive analytics and data mining. The key word that bothers civil libertarians is "predictive," as it suggests that police and other national security agencies are gathering data on people's behavior even though they have not committed a crime. As Eligon and Williams note, the Chicago police have developed a "heat list" of 400 people who are considered more likely than the average citizen to become engaged in a violent crime. The economic, political, and social stresses that will accompany the breakdown in natural systems, and the civil wars fought over the increasing scarcity of water and other resources, will lead to greater police efforts to assert central control, especially over the critics of government policies that favor the rich and politically powerful. Criminality, as we witnessed in the fascist regimes of the last century, can be redefined to include those who challenge the fascist drift of the country. Predictive analytics is already leading to a list of professors regarded as radical and thus subversive, including professors who are thought to be sympathetic to the idea of a Palestinian State.

Increased reliance upon computer-mediated learning at all levels of education contributes to the conformity of thinking needed in a techno-fascist state. Lost are the ethnically diverse intergenerational narratives passed forward through face-to-face relationships, which leaves students exposed only to myths that serve the interests of the controlling elites:

scientists, computer scientists/engineers, corporate heads, and a military establishment increasingly concerned that the ecological crisis will disrupt its hegemonic agenda. The guiding ideology and moral codes were first articulated in the early 17th century by Johannes Kepler, who suggested that life processes should be understood as machine-like. This notion continues to be reinforced by computer scientists who have announced the beginning of the post-biological world—and their followers—who rely on the values of efficiency, accountability, profit, data, and purposive rationality to engineer machines that replicate human behaviors and thought processes (their own, of course).

Learning programs reflect the presuppposed world of the people who write the software. This presupposed world already includes the practice of relying upon print and data to communicate the false sense of a factual and objective account of social events and ideas (as though the interpretative frameworks of writer and reader have no influence on what is written or represented as data). Given the above, there is less chance that students will recognize that the words appearing in print are metaphors encoding the thought processes of earlier generations—people who were unaware of environmental limits. In short, conformity of thinking relies upon (i) a mix of mythologies, (ii) the elimination of historical/cultural memories that do not support these guiding myths, and (iii) vocabularies that limit thought and thus awareness that the technological/economic/industrial/military elites have set an agenda for everyone to follow.

Questions that need to be asked about parallels between the European varieties of fascism and American rightwing groups include the following: Is there a parallel between how the German National Socialists in the 1930s manipulated democratic process to gain support of their totalitarian agenda and how the U.S. National Rifle Association uses the protection of the Constitution and congressional campaign contributions to support its agenda of arming rightwing hyper-patriotic Americans who stand ready to intervene should a peace movement break out across the country? What about the parallels between the male-dominated fascist movements in Europe and the male-dominated fields of computer science, engineering, national security agencies, the military establishment, and corporations whose future is tied to the digital revolution? Does a concern with data, efficiency, and a vision of progress easily interpreted in the language of Social Darwinism reflect the West's deep assumption that this is a human-centered universe which should be guided by the scientific rationalism of men? Are the roots of violence, especially toward women, traceable to the monotheistic religions in the West, where God is associated with masculine (so interpreted) qualities?

But fascism also relies upon the combination of conformity in thought and values, of loss of historical memory, and of perceived crisis or endpoint that requires the collective energy and loyalty of young and old. In addition, there needs to be a significant percentage of the population that is hyper-patriotic, thinks in clichés, and is willing to support the use of imprisonment and torture of those who challenge the rise of techno-fascism—especially those labeled as environmentalists—who will be viewed, as were the Jews in Nazi Germany, as weakening the power of the state.

Digitally mediated learning, which is heavily dependent upon print and data-based accounts that encode the cultural assumptions and ideologies of the people who write the programs, reinforces a mindset that responds to short explanations. Already, boredom is associated with long explanations and written accounts. The ways in which social media reinforces the importance of an ever-shifting sense of immediacy and instant response to anonymous Others ensures that the emergence of a fascist state will go unrecognized. The increased reliance on mobile devices has led to the coining of a new word, "micromoment." This is now understood by the cutting-edge advertising industry as the current length of time they have to bring a product or service to the attention of a potential consumer. The understanding within the industry is that a potential consumer's attention span is eight seconds. The ultimate goal is to get the information on a product or service to appear on a mobile device the moment it occurs to the potential customer as a need.

The shortening of attention spans and the expectation that there is an instant answer to fulfill personal needs should be understood within the larger context that takes account of impacts on natural systems, historical memories of social justice traditions, the growing disparities between vast numbers living in poverty and the super-rich, and the social tensions that will rise exponentially as environmentalists challenge those who have an economic and ideological interest in denying the existence of the ecological crisis. Those in denial have a vested interest in how techno-fascism continues to shape human consciousness to equate progress with even higher levels of efficiency, personal conveniences, and allowing constant monitoring by medical experts of one's bodily functioning and appearance.

The totalitarian nature of the digital revolution can be seen in the proliferation of apps that represents both a hyper-growth industry and a personal quest for quick riches on the part of younger people who are part of the growing contingent workforce (now a third of employed Americans). They must rely upon their own economic efforts now that the old system of lifelong employment, with its broken system of social contracts, is fast disappearing. The drive is to compete with others in creating an app that attracts so many users that it leads to great wealth and thus banishes the anxiety of becoming economically destitute in later years. After all, safety nets cannot be justified in the monoculture emerging from the integration of techno-fascism and libertarian/market liberalism. This new growth market, like the gold rush in California, does not draw the reflective and ecologically informed, but rather those who want to develop the app that will be the panacea for those who want to get on the technology bandwagon. The ways in which techno-fascism is overwhelming individual, historically and even ecologically informed judgments can be seen in the number of apps for improving math instruction, reading, teaching in other curricular areas, and classroom management. There are more than 3900 school-related apps, so many that school districts now rely upon external experts to guide them on the merits and uses of apps.

Nearly every aspect of personal and social life can now be organized and monitored by anonymous experts who lack any historical knowledge of how to live a more ecologically sustainable and community-centered life. Deep knowledge of local contexts, cultural differences, intergenerational knowledge, the importance of personal memory, critical judgment, and wisdom traditions from the past, are not part of the life-world of the generation raised by spending hours in the abstract and individually centered world of digital culture. Like other forms of cultural reproduction, they simply reproduce the taken-for-granted way progress is understood within the emergent techno-fascist culture that is now being celebrated (i) in the media, (ii) in people's economic behaviors, (iii) by politicians supporting the NRA, and (iv) by the mega-corporations that control major segments of American life and are major forces behind the denial that there is an ecological crisis.

The face-to-face systems of local control involving a variety of democratic practices and traditions of ecological wisdom, and which often involve religious and indigenous communities, are under threat from (i) the abstract knowledge read on a computer screen, and (ii) from the myth of individual decision-making in the unlimited world of data. Where in the digitally mediated curriculum will students learn about the ecologically sustainable traditions of their communities? Where will they learn of past mistakes? The ideology underlying the digital revolution represents traditions, including local decision-making, as a source of backwardness and an impediment to students creating their own ideas from the wealth of context-free data available on the Internet.

It is even becoming difficult to recognize the differences between local activism focused on social justice issues and the way corporations such as Coca-Cola, Wal Mart, Chick-fil-A, and health insurers (organizers of the "town hall" protests against President Obama's health care legislation) now use social media to engage in what is called "astroturfing." This is the term used to distinguish between genuine grassroots social justice movements and the supposedly "grassroots" groups created by corporations to advance their economic agendas. Corporations regularly deflect criticism of their harmful products by using pseudo science and social media to create the impression that a wide segment of the public supports them. For example, Coca Cola has spent millions funding the Global Energy Balance Network, which is headed by scientists who claim that lack of exercise and not the sugar content of soft drinks is the cause of obesity. Putting profits above the well-being of the public is yet another way in which many American corporations emulate German corporations in the 1930s that supported the transition from a weakened democracy to the militant and expansionist vision of National Socialism under Hitler. Indeed, a number of prominent American corporations such as

Ford, Kodak, Coca Cola, and Standard Oil were essential to the German war effort, along with BMW, Siemens, and Bayer. Many American youth now identify the last three named as American companies.

Digital technologies now have indispensable uses that range across a wide range of cultural activities, from medicine, scientific research, monitoring and maintaining the society's technological and economic infrastructure, education, and nearly every facet of the industrial/consumer-dependent culture. But digital technologies have also introduced irreversible cultural changes, such as undermining local democracy (did we vote for any of these technologies?), creating a new generation that is unaware of the political dangers and threats to personal security that accompany the loss of privacy. They have undermined the (i) face-to-face intergenerational narratives essential to maintaining ethnic identities and (ii) traditions of the cultural commons that strengthen patterns of mutual support while reducing dependency upon consumerism. They have further strengthened the long-standing tradition in the West of elevating abstract knowledge over ecologically informed ways of thinking. Digital media is shortening people's attention span to the point where little more than slogans and sound bites serve as the basis of political decision-making. Following the standard set by Fox News of masking disinformation as a model of factual accuracy and objective reporting, millions of Americans have been conditioned to accept ideologically driven propaganda, which further reduces the likelihood of mass resistance to the techno-fascist agenda.

The fashion industry has never been on the side of conserving natural resources, but has always been in the forefront of pushing high-status, wealth-driven images of how successful people should dress. As both the design and use of materials promoted by the fashion industry send powerful messages about what should be valued and thus emulated within the larger society, the growing number of human-like robots has inspired fashion trendsetters to dress humans in ways that suggest the older, more mechanistic appearing robots. With the line that separates humans from robots becoming increasingly blurred, the message in the futurist film, *Ex Machina,* which might have sparked resistance to the computer scientists' understanding of progress, is likely to be lost. The comely female-bot outwits the highly gifted employee of a software company and then disappears into the flow of street traffic without leaving a hint of her non-biological origins. The robot was not only more beautiful but more intelligent and crafty than the humans who created her.

The critical question is whether there will be resistance to how everyday lives are being increasingly monitored, motivated to pursue the increasingly narrow economic agenda of the emerging techno-fascist culture, and stripped of historical values and identity. Will enough of the public recognize the dangers that lie ahead? Will they be able to articulate the importance of what is being lost? It is important to note that the computer scientists who play a central role in articulating the ideology that underlies the emerging techno-fascist culture totally ignore the deepening ecological crisis. The title of the book written by Peter Diamandis and Stephen Kotler, *Abundance: The Future is Better than You Think* (2012) could serve as the anthem as we march into the future envisioned by techno-fascists. It bears repeating: the scientific justification for replacing humans and their diverse cultures with the culture created by super-intelligent computers, according to a number of computer scientists following the lead of Ray Kurzweil (1999, 2005, 2012), is simply the process of natural selection.

Traditional defenses against totalitarian regimes are now being lost. To understand this, we need to focus more specifically on the cultural transformations that occur in the classroom. Here, students increasingly spend more of their day in computer-mediated learning, which displaces face-to-face interaction with teachers who might spark their curiosity to explore beyond the orthodoxies of the day. The many hours daily spent texting friends, playing video games, and exploring the seemingly endless realms of cyberspace also shortens attention spans in ways that undermine long-term memory. Speed and context-free slogans have now replaced depth of understanding and critical judgment. The same shift to computer-mediated thinking and communication is occurring in work settings, and in nearly every other venue in society—including vacation time. The current business ethos is for managers to constantly monitor how employees spend their time, including how quickly they perform different tasks. The panoptican system that Jeremy Bentham designed for keeping prisoners under constant surveillance is now being electronically extended to the office and other workplaces.

A brief review of previous chapters will bring out the cultural changes undermining democracy and ethnic difference, changes that are leading to the monoculture of digital consciousness.

Print, Data, and Abstract Thinking

Print is unable to represent the emergent, relational, and co-dependent nature of biological and cultural ecologies, whether it is the multiple message systems being communicated as one walks through an open meadow, or interaction with honors students who argue that their ideas are original even though they have read a number of articles explaining how the analogs that frame the meaning of words they take for granted were settled upon in earlier eras. A printed account of the emergent, relational, and co-dependent behaviors in both the natural and cultural ecologies, especially if they are personal experiences, would provide only a surface and static (and thus abstract) account. What gets recorded in print is also influenced by the writer's biases, presupposed ways of understanding, and interpretative frameworks. These will all go unrecognized due to past socialization. The printed (also spoken) word relies upon a conduit (sender/receiver) view of language, which too often leads to ignoring that words have a history that influences what the reader will or will not be aware of.

The printed word, data, and oral communication that is influenced by the abstractions of the printed word shift attention away from the ongoing, embodied, and face-to-face experiences with others in the community. We are conditioned to accept as real the abstract accounts of what is happening in society. Bombarded daily from the media with threats and proposals for change, we become indifferent about assessing the accuracy of these claims and are thus easily led to think and value what fits the agenda of the controlling elites. The printed word, in effect, diverts attention away from the relational, emergent, and co-dependent nature of face-to-face and culturally mediated experiences.

For those who promote the worldview of a totally digitized future, the mythic thinking that merges the idea of progress with the meta-narrative of natural selection enables them to represent themselves as oracles of Nature's and the culture's destiny. Both become powerful sources of authority, especially when people's privacy, employment, sense of history, and cultural identity are monitored by the flood of sensors being introduced into every aspect of daily life.

The political importance of "connectivity," which computer scientists view as a strength of digital technologies, is largely unrecognized by the general public. Without it, the massive amount of data gathered on people's lives would not be so widely distributed to the various agencies—from producers of goods and services to the National Security Agency and other governmental agencies.

An ideology that drives the unrelenting quest to digitize every aspect of human experience represents the future as leading to improvements over the past. A state of consciousness wherein the expectation of a better future becomes dominant leads in turn to ignoring (i) traditions that are sources of everyday empowerment, and (ii) hard-won institutionalized ways of protecting one's civil liberties. When the past is no longer viewed as a mix of prejudices and social justice achievements, the depth and complexity of everyday experiences becomes narrowed and simplified by the limitations of what can be represented in print and digitized forms.

As the Internet of Things leads to the wireless connectivity of all aspects of people's personal lives with everything else in the environment, the experience of being constantly watched may lead to a sense of security. But even small incidents should be taken as a warning that computer scientists have created, all in the name of progress, an interconnected policing system. A case in point: a typing error in one number of my house address by one agency led to the same error being repeated in mailings from a variety of governmental agencies and corporations wanting to promote their products. Now that this kind of connectivity is increasingly being used to anticipate acts of terrorism, or where crimes will occur, perhaps it is time to stop referring to this as surveillance and call it what it is: a policing system.

The next step will be to monitor potential sources of dissent—a problem that scientists are now working on as they study the connections between people's vocabularies and their patterns of thinking. Other scientists are making progress along the same pathway pioneered by Nazi scientists in developing facial recognition technologies. There are current efforts to discover the chemical changes needed to eliminate bad memories, which, then, can also be used to eliminate good memories (such as privacy and a life free of commercialism). Other efforts range from trying to extend Lee M. Silver's technologies for bio-engineering babies with traits wanted by their parents to engineering the conceptually and morally compliant babies needed in a techno-fascist state.

One of the ways that techno-fascism spreads throughout society like an unrecognized virus is that its supporting language relies heavily upon metaphors that are reassuring to many people. Most Americans support "progress," "security," "conservatism," "Americanism," and "patriotism." When they lack the conceptual background (thanks to omissions in our educational systems) they are unlikely to recognize that the so-called conservatives are actually a mix of libertarian and market liberals, and that the above metaphors have been used to justify imperialist-driven wars.

The expansion of surveillance of people's lives adds another layer to the fascist political agenda of the American rightwing that mirrors key characteristics of European fascism. This social agenda includes (i) placing barriers in people's ability to vote, (ii) using the prison system to control a large segment of the poor and non-white population, (iii) the intertwining of fundamentalist religions and segments of the government focused on national security, (iv) using the military to globalize the American way, (v) suppressing basic human rights, especially for women, (vi) undermining the rights of workers to organize, and (vii) enabling fraudulent elections where the super-wealthy are able to control the outcome of state and federal elections.

The expansion of technological and corporate power involves greater reliance upon the use of context-free metaphors such as "national security" and "terrorist" to justify using the power of the police against individual and social groups engaged in demonstrations and acts of resistance against corporations that continue to have a destructive impact on the environment and peoples' lives. As sources of protein become even more limited due to the warming and acidification of the oceans, as droughts reduce the viability of genetically engineered seeds (and companies such as Monsanto reduce the variety of seeds in order to increase farmers' dependence), and many other scenarios that will be played out as ecological systems collapse, greater social unrest will occur in response to a variety of issues that the money-controlled state and federal governments have not addressed.

Social unrest will be further exacerbated as new digital technologies continue to reduce the need for workers, and as youth remain unaware of the political economy of the cultural commons where talents and patterns of mutual support lead to non-monetized forms of wealth. This is when the all-encompassing digital infrastructure will be used to suppress all forms of dissent—while at the same time relying upon metaphors that suggest police state tactics are intended to provide security in an increasingly chaotic world. As the deepening ecological crisis and the displacement of humans by machines increases, techno-fascism will become the new normal.

Online Courses and Degrees

Changes are occurring so rapidly that we seem unable to connect the dots. Recent insights into how language issues are related to the ecological crisis have not yet become a widespread concern, with the result that the mythical thinking encoded in many of our guiding metaphors still goes unchallenged. When only the important, indeed indispensable, uses of print and data are understood, then little attention is given to how the merging of these technologies is changing the deep conceptual foundations of consciousness.

Also, the understanding of fascism by most Americans is limited to the short newsreel images of the World War II era—if even that. Few understand that it is a modernizing ideology, and that it relies upon myths. These now include yet another Social Darwinian interpretation that emphasizes, according to computer futurist writers Gregory Stock, Ray Kurzweil, and George Dyson, the transition to a post-biological world governed by super-intelligent computers.

Given this widely shared state of consciousness, promoting an even greater reliance upon using computers at all levels of public and higher education is understood as necessary to progress—even as digital technologies are radically changing the prospects of earning a living and taking for granted that the past has been made totally irrelevant. As soon as the pre-online generations die off, what remains of historical memory will depend upon the programmers' and data scientist's taken-for-granted world, which will be restricted by the way their field of graduate study excluded the need to understand the complexity of the cultures into which their digital technologies are introduced.

The dots we now need to connect include what will *not* be learned online, and how that in turn will contribute to (i) not being able to recognize the cultural/linguistic roots of the ecological crisis, (ii) not knowing alternatives to the near total dependence on a consumer lifestyle that is leading to ecological collapse, and (iii) not recognizing that the migration of millions of people from areas no longer able to support life will be magnified many times over as extreme weather, rising ocean levels, resource wars, and technologies further displace the need for workers.

Online courses and now online degrees offered by universities, ranging from the ever-present degree mills to elite universities, have many advantages. They are more convenient in terms of meeting a wider range of student income and family needs. And the need to attract students has led to aligning course requirements more closely to different career paths—which means that technical knowledge is prioritized over courses that would provide students a more critically informed historical and cultural perspective.

Perhaps the most important source of empowerment essential to resisting the conformity required by a techno-fascist regime is the ability to recognize and reframe, in terms of current cultural and ecological understandings, the many forms of metaphorical thinking that reproduce the assumptions and silences of earlier eras. Understanding and, better yet, participating in the mutually supportive lifestyles of the cultural commons represents ways of resisting the environmentally destructive commercialism that is central to techno-fascism. In effect, online education repeats the shortcomings of bricks-and-mortar public schools and universities.

School gardens and recycling practices fall far short of the cultural issues classroom teachers should be addressing, particularly since students in the early grades will be living, in 50 years or so, through the early stages of social unrest that will turn more violent in an already over-militarized world. Similarly, online university courses and degrees, still driven by the hubris that has become endemic among academics focused on overturning all traditions even though they are unaware of how many of them are sources of personal empowerment, will continue to perpetuate the Ponzi scheme of keeping up enrollments by promising to prepare students for careers even as the careers are being taken over by algorithms and robots.

If one only considers how public schools and most aspects of higher education continue to ignore Albert Einstein's warning about using the same mindset to correct today's problems that was used in creating them, as well as how far down the pathway of techno-fascism we have already travelled, it would seem there is little hope for the future. But as suggested in earlier chapters, the hope for the future can be seen in the cultural commons activities that need to become more widely recognized as sites of resistance, as well as community-centered lifestyles that rely upon democratic decision-making. These face-to-face communities are also less prone to electronic surveillance and thus less dependent upon the technologies essential to the control systems required in a techno-fascist state.

The next chapter will address the grass roots alternatives being explored in different regions of the world for passing forward the inter-generational knowledge and skills essential to strengthening the cultural commons—that now go by such different labels as the localism movement, transition communities, cooperative movements, and the revitalization of indigenous identities and cultures.

References

Dyson, G. 1998. *Darwin Among the Machines: The Evolution of Global Intelligence.* New York: Basic Books.

Eligon, J, and J. Williams. "On Police Radar for Crimes They Might Commit." *New York Times,* 25 September, 2015.

Kurzweil, R. 1999. *The Age of Spiritual Machines: When Computers Exceed Human Intelligence.* New York: Viking.

Stock, G. 1993. *Metaman: The Merging of Humans and Machines in a Global Superorganism.* New York. Doubleday.

Index